THE ROYAL AGRICULTURAL SOCIETY OF NATAL, 1984–2021

Bill Guest

Natal Society Foundation and Royal Agricultural Society of Natal
PIETERMARITZBURG
2022

Cover

The Royal Agricultural Society's coat of arms symbolically depicts the influences that contributed to its history. The supporter on the left is a unicorn taken from the coat of arms of the United Kingdom. Its lance symbolises the role played by the British Army in the history of Pietermaritzburg and the Royal Show.

The supporter on the right is a wildebeest taken from the then South African and Natal coats of arms, the assegai being symbolic of the Zulu nation. The three rings in the top portion of the shield represent the three showgrounds occupied by the Society since the first Show in 1851. The wavy line separating the top of the shield from the lower portion represents the Dorpspruit, which flows through the current premises. The elephant in the lower portion was taken from the Pietermaritzburg coat of arms.

The red rose surmounting the crest refers to the City of Flowers and to the roses that apparently grew prolifically in Pietermaritzburg's hedgerows in 1851. The shield and supporters stand on a base of green grass with a white railing in the background similar to the fence that surrounded the main arena.

The Latin motto 'Agriculture is the strength of the State' was adopted in consultation with Alexander Petrie, professor of Classics at the University of Natal. This coat of arms was registered in 1975.

This book is dedicated to all those who have contributed in various ways to the ongoing success of the Royal Agricultural Society of Natal and its activities

CONTENTS

AUTHOR'S NOTE

The use of racial terminology is always controversial and potentially offensive. This is particularly the case in the South African context in which, unfortunately, its extensive presence in historical commentaries and records makes it impossible to avoid. For the sake of consistency, in this book those references that imply a place or country of origin are given in upper case, for example, African, Afrikaner, European and Indian. References that denote dubious classification by skin colour are given in lower case, for example, black, coloured and white. No offence is intended in the use of these terms.

ABBREVIATIONS

ABI	Amalgamated Beverage Industries
AGRI	South African Agricultural Union (Agricultural Society of SA)
AR	annual report(s)
BBEE	Broad-Based Black Economic Empowerment
BEE	Black Economic Empowerment
CCMA	Commission for Conciliation, Mediation and Arbitration
CCTV	closed circuit television
CEO	chief executive officer
COSATU	Congress of South African Trade Unions
CPIX	consumer price index
DIY	do-it-yourself
DUT	Durban University of Technology
ESKOM	Electricity Supply Commission
EXCO	executive committee (RAS)
FEI	Fédération Equestre Internationale
FM	frequency modulation
FMX Trains	freestyle motocross
FNB	First National Bank
ICC	International Convention Centre (Durban)
ID	identity document
JCB	J.C. Bamford Excavators Limited
JSE	Johannesburg Stock Exchange
KFC	Kentucky Fried Chicken
KWANALU	KwaZulu-Natal Landbou-Unie (Agricultural Union)
KZN	KwaZulu-Natal
MANCO	management committee (RAS)
MEC	member of the provincial executive council
MUHL	MUHLE=Outdo
NAMPO	South African Agricultural Trade Show
NASREC	National Recreational (Expo) Centre
NCD	National Co-operative Dairies
n.d.	no date
NEHAWU	National Education, Health and Allied Workers' Union
NFA	Natal Field Artillery
NGO	non-governmental organisation
no.	number
NPA	Natal Provincial Administration
OMI	Order of Mary Immaculate (Catholic)
PADCA	Pietermaritzburg and District Care of the Aged
PCB	Pietermaritzburg Chamber of Business
PMB	Pietermaritzburg

RAS	Royal Agricultural Society
RASC	Royal Agricultural Society of the Commonwealth
RPO	Red Meat Producers Organisation
RSG	Radio Sonder Grense
SAAF	South African Air Force
SABC	South African Broadcasting Corporation
SACP	South African Communist Party
SAMIC	South African Meat Industries Company
SAMRO	South African Music Rights Organisation
SANDF	South African National Defence Force
SAP	South African Police
SAPS	South African Police Service
SARS	South African Revenue Service
SASSA	South African Social Security Agency
SATV	South African Television
SPCA	Society for the Prevention of Cruelty to Animals
SWOT	strengths, weaknesses, opportunities and threats
UCI	Union Cycliste Internationale
UKZN	University of KwaZulu-Natal
UNISA	University of South Africa
USA	United States of America

LIST OF ILLUSTRATIONS

FOREWORD

At times referred to as the 'grand old lady of Pietermaritzburg', 23 December 2021 will see the Royal Agricultural Society of Natal celebrate its 170th anniversary. Few South African institutions can lay claim to such a proud achievement. The Society's evolvement has been largely overseen by ladies and gentlemen who have magnanimously given of their time to grow the organisation from a local agricultural event to a facility hosting the largest and most respected mixed exhibition in the country today; the Royal Show. As an adjunct, the well-maintained Showgrounds and its structures have become increasingly popular for functions and events, including the prestigious annual opening of the provincial legislature.

It hasn't always been plain sailing, and by way of the exhilaration accompanying expansion and progress, the Society has also been confronted by its own share of challenges. These include the horror of the rinderpest outbreak of 1897, the Anglo-Boer War, the Depression, two world wars, and very recently, the social and economic dislocations associated with Covid-19. Regrettably, these occurrences have necessitated the cancellation of the Royal Show on eleven occasions, including those of 2020 and 2021.

More recently, the pressures of urbanisation have impacted the Society's operational capacity. We were the 'first kid on the block' in 1902, but our current and much-loved venue is no longer fit-for-purpose. Having served as home for the past 119 years, logistics have become increasingly strained, while the incongruity of catering for several thousand livestock in the centre of a commercial district speaks for itself. This will, in the near future, necessitate relocation; sad for many, but also a new beginning.

Any negatives notwithstanding, the resilience of the Society has never been found wanting, and in each instance, the RAS has commendably risen to the occasion. All-in-all, a formidable and proud record warrants this third update of the Society's history; this one spanning the years 1984 to 2021.

To Professor Bill Guest who has ably undertaken this exercise, to all persons who contributed to the publication, to the Natal Society Foundation, and most importantly, to the many unsung heroes who have overseen the Society's development over the past 170 years – your perseverance and generosity of spirit is acknowledged with praise and approbation.

T.D. Strachan
CEO, Royal Agricultural Society of Natal
June 2021

PREFACE

It is a great pleasure and privilege to have been entrusted with the task of updating the history of one of KwaZulu-Natal's oldest and most valued institutions. Among several others these include the art galleries, hospitals, libraries, municipal facilities, museums, schools and tertiary colleges that were established by those far-seeing residents of the nineteenth and early twentieth century. In so doing, they contributed so much to the cultural heritage of this region that we now take for granted.

This book is based on the firm foundations laid by Lindsay Young's *A History of the Royal Agricultural Society of Natal, 1851–1953* (Pietermaritzburg: Royal Agricultural Society, 1953) and Ruth E. Gordon's *Natal's Royal Show* (Pietermaritzburg: Shuter & Shooter, 1984). Their work is condensed in the Prologue but should be read in the original form in order to be fully appreciated. The words highlighted in bold text in the chapters that follow are intended to facilitate ready access to those aspects of the Society's history in which the reader may have a particular interest without the necessity of referring to the index.

Thankfully, the Royal Agricultural Society of Natal has never forgotten its agricultural roots, even though the Royal Show is no longer the only significant event on its annual calendar. It was the then very prominent agricultural dimension of the Show that first captured my interest as a schoolboy. In the 1950s my classmates and I were bussed up from Durban to enjoy the sights, smells and sounds of farm animals and equipment which was so different to the environment with which we were familiar as townies.

A further attraction was provided by the gaggles of giggling gerties with their various unfamiliar Midlands school uniforms and their pink cheeks that were so different to the tanned damsels of the beachfront. We were particularly awed by the show jumping; not least by the junior contestants, one of whom fell heavily in the main arena before our very eyes, but remounted to complete her round amid thunderous applause. It may have been the girl I was to take a fall for and marry many years later in Pietermaritzburg!

As a resident of that city from the late 1970s I became a regular visitor to the Royal Show and the funfair, in the company of my first (late) wife Denise and our children Annabelle and Philippa. A friendship with my wife's Estcourt schoolmate, the now late William Dreboldt and his spouse, former juvenile show jumper Cynthia, served to reinforce our interest and knowledge of the

Royal Agricultural Society and its Show thanks to their enthusiasm. William was for me the personification of those many voluntary helpers, staff members, exhibitors and competitors who over the years have served the Society in so many different ways. It is therefore to all of them that this book is dedicated.

My sincere thanks are due to those persons listed in the Bibliography who shared with me their reminiscences of the Royal Agricultural Society of Natal and its various activities. They all did so with such encouraging enthusiasm. I am particularly grateful to Dr Iona Stewart for suggesting, on Jack Frost's recommendation, that I write this history; and to her, and the Society's CEO Terry Strachan, for valuable information.

I also wish to thank Janice Will and Irene Peters, who so efficiently assisted me in locating the relevant records in the official archives. Jo Marwick undertook the onerous task of design and page layout.

Not least, I am hugely indebted to my wife Cynthia (formerly Dreboldt) for encouraging me in pursuing this worthy project and providing her computer expertise in moments of crisis to ensure its completion.

W.R. (Bill) Guest
December 2021

PROLOGUE

THE COLONIAL ERA in KwaZulu-Natal gave birth, among numerous other local institutions, to the Royal Agricultural Society (RAS) and its annual Royal Show. What was initially known as the Pietermaritzburg Agricultural Society was established in 1851. This was little more than a decade after the Voortrekkers had defeated King Dingane kaSenzangakhona's Zulu kingdom at Ncome (Blood) River in December 1838 and the following year had declared what was briefly known as the Republic of Natalia. In 1842 it, in turn, had been invaded and subsequently annexed as an autonomous district of the distant British Cape Colony.

The ill-starred Boer leader Piet Retief was credited with having chosen the well-watered site for the Republic's administrative capital, Pietermaritzburg. There, a palisaded camp of mud huts and wagons soon gave way to brick and stone dwellings served by water furrows supplied from the smaller Boesmansrivier (Dorpspruit) and the larger Boesmansrivier (Msunduzi).

Following British annexation most of the 4 000 trekkers retreated back across the Drakensberg in favour of republican independence, leaving barely 100 families behind. From the 1850s they were joined in the new Colony by an influx of mostly middle-class British settlers. The new arrivals were driven by recessionary conditions at home and attracted by the often-glowing descriptions of Natal that various immigration schemes disseminated. Some of the settlers were smallholder farmers lured by the opportunity, now virtually impossible in England, to become large-scale landowners.

During the nineteenth century far greater immigrant numbers were attracted to North America and to Australasia than to southern Africa. By the early 1850s Natal's settler population amounted to little more than 7 600, rising to barely 18 000 by the late 1860s amid an estimated indigenous population of more than 250 000.

By 1854 one third of the Colony's new British arrivals, including some of those with prior farming experience, had already drifted off their allocated plots into the towns in search of alternative employment or entrepreneurial opportunities and a more congenial lifestyle. They had found it impossible to cope with unfamiliar climatic conditions, insufficient land and working capital, a sense of isolation and, in some cases, inadequate access to water and markets. New urban centres like Greytown, Howick, Richmond, Verulam and York emerged; but there were initially only two of any significant size.

Pietermaritzburg quickly attracted 1 500 white residents but Port Natal soon emerged as the region's major economic growth point. British traders had established themselves there as early as 1824 and in 1835 had renamed their outpost Durban in an effort to gain the support of Cape Governor Sir Benjamin D'Urban for British annexation. By the mid-1870s at least half of Natal's white population already resided in these two urban centres. There and elsewhere their increasing numbers and economic activity were having a widening impact on the lives of the indigenous population as well as on the flora and fauna of the region.

The grid-pattern layout the Boers had brought to Pietermaritzburg comprised eight untarred trekker-named streets that formed blocks parallel to Church Street on the town's central ridge. This ran from elevated ground to the west, where the British garrison's Fort Napier was established, down to the Dorpspruit vlei in the east. Roughly half-way the Church of the Vow, adjacent Market Square and Raadzaal formed the spiritual, socio-economic and political hub of the town.

Initially, erven within the grid were virtually smallholdings of about 0.8 hectares in size on which crops and orchards could still be cultivated, unlike the more familiar subdivided stands of today, while livestock was pastured on townlands. Each street intersected Commercial Road (previously Nelstraat and now Chief Albert Luthuli Road) at right angles as it ran through town from the bottom of the hills to the north, down towards the Msunduzi before leading on to the rough wagon track beyond it to Port Natal/Durban.

Pietermaritzburg acquired the status of a city in 1854 when J.W. Colenso's arrival as Bishop of Natal elevated it to an episcopal see. In 1856 it officially became a colonial capital when Natal was established as a separate colony from the Cape, with representative government granted to the white settlers. Its own supreme court followed in 1858.

In common with settler communities elsewhere during the nineteenth and early twentieth century Pietermaritzburg's residents consciously sought to improve their quality of life by establishing institutions in imitation of the society they had left behind. Theatrical and turf club events were held from 1844 and a Reading Society was started in 1845 to promote the written word and the English language that was so central to settler culture. In 1846 David Dale Buchanan launched the *Natal Witness,* preceded by Cornelius Moll's and Charles Etiennne Boniface's short-lived *De Natalier* two years previously.

The first coffee house appeared in 1847 and in May 1851 the Natal Society was established. It assumed responsibility for the Reading Society's library and subsequently developed it into a major public facility, now known as the Bessie Head Municipal Library, while also promoting monthly musical recitals.

Grey's Hospital opened in 1855 and the London-style Victoria Club (now the Victoria Country Club) in 1859. From 1864 St George's Garrison Theatre at Fort Napier began to rival the productions staged downtown while the performances of the regimental band were always much appreciated. Indeed, the presence of a resident imperial garrison until as late as 1914 contributed significantly to the cultural and social life of the city and also to the local economy.

The rich heritage the settler community left for posterity eventually also included a variety of churches representing different Christian denominations, a city hall and council, a municipal administration, electrical lighting, water and sewerage systems, banking institutions, a chamber of commerce, sports clubs and other public amenities.

Among these were Alexandra Park (1863), the Botanic Society and the Natal Law Society (both established in the 1870s), the Tatham Art Gallery (1903) and the Voortrekker (now Msunduzi-Ncome) Museum (1912). The settlers also founded several schools and tertiary educational institutions including the (now Msunduzi) Technical College, a Teachers' Training College (1909) and the Natal University College (1909).

Yet for all the trappings of nineteenth-century British urban sophistication, Pietermaritzburg was still a market town at the heart of a predominantly agriculturally based economy. By the early 1870s nearly two thirds of its residents were still engaged in agriculturally related enterprises. There was little industry in the absence of sufficient investment capital and skilled labour. For much of that century manufacturing development was concentrated primarily around the harbour city of Durban.

The *Natal Witness Express* provided an invaluable postal service between the two centres with runners who initially took two days to complete the journey compared with the three or four days by ox-wagon. From 1860 both were eclipsed by the six-horse, twelve-passenger omnibus that took a mere eleven hours, followed in 1864 by a privately owned telegraph line. The military authorities assisted hugely in improving the road link and from the late 1850s the Mngeni and Msunduzi rivers were both spanned by iron tension bridges.

Initially, little was known about the broader geographical details, bioclimatic variations and agricultural possibilities of the region. Early maps of Natal-Zululand were unreliable, even with regard to major characteristics such as the exact situation and extent of the Drakensberg range and of the primary rivers flowing from it.

The public library that was to make the Natal Society so well-known was not its initial purpose. It was rather to assist the farming community to develop the agricultural potential of Natal and to disseminate accurate information that would consolidate it as a British colony by attracting more immigrants. Indeed, its draft rules of 17 June 1851 specifically focused on the need for more details about the region's physical attributes, soil, climate and indigenous population. They also called for proposals as to how these might best be developed. To that end during the first five years of its existence the Natal Society held nearly twenty public lectures and appealed to local farmers to contribute information relevant to these issues. Enterprising individuals were already experimenting to identify the most viable forms of farming within the region's dramatically different bioclimatic zones as the topography rose steeply between the coast and the mountains.

After setbacks in the production of arrowroot, indigo, cotton, rice and coffee they enjoyed some success with tea, but eventually settled primarily for sugar along the coastal strip. Wool, maize, wattle production and dairy cattle were found to be most appropriate in the Midlands while mixed farming, including sheep grazing, cattle ranching, fodder crops, maize and later cotton and wattle, was best suited to the northern districts.

The Natal Society was involved in two further significant initiatives. Its launch of the Natal (now KwaZulu-Natal) Museum in 1904 was delayed by limited funding. Its close association with the town's Agricultural Society was yet another indication of its resolve to advance what was still a predominantly agricultural economy.[1]

The Pietermaritzburg Agricultural Society (1851–1904) held its first meeting in November 1851 on the premises of the Natal Society with which it initially shared the same office bearers and committee members. In the second edition of his *Natal Witness* in March 1846 David Dale Buchanan had questioned why the Colony did not have an Agricultural Society. The Cape had established one in 1831 and in 1848 Durban followed suit with its Natal

Agricultural and Horticultural Society.

The latter's focus was increasingly on the development of that city's Botanic Gardens whereas the Pietermaritzburg Society concentrated on the promotion of agriculture by mounting exhibitions. Displays were very much in the tradition of the Cape Society, England's Royal Agricultural Society (1839) and the Great Exhibition held at London's Crystal Palace in May 1851.

Pietermaritzburg's first Show was held on 23 December that year in the present-day town gardens, below where the Presbyterian Church now stands and across the road from the Raadzaal (now City Hall) site on Commercial Road. It was dampened by summer rain and in 1852 two shows were held, in June and at Christmas, the latter including a ploughing competition beside the Msunduzi.

From 1854 there was only one annual Show, in the more predictably drier month of May and on the nearby Market Square, just east of the City Hall. This remained the venue, without any municipal charge, for 35 years with the mayor being an ex officio member of the Society's committee.

The Agricultural Society, like the Natal Society, was initially desperately short of funds. Increased subscriptions and donated prizes improved the situation as the small white farming community gradually found its feet. However, the depression of the 1860s, along with insect infestations, sheep scab, lung sickness, redwater fever and rising freight costs all impacted upon farmers' and the Society's finances. So too did the discovery of diamonds in the interior, which attracted several young farmers to the Kimberley region and induced others to switch to full-time transport riding.

The 1873 Langalibalele crisis, followed by the 1879 and 1881 British defeats at Isandhlwana and Majuba at the hands of the neighbouring Zulu and Boers respectively, further undermined confidence in the Colony's future and that of their Agricultural Society among isolated farmers. The existence of other, smaller agricultural societies in Natal, twelve by 1895, promoted competition but was a dissipation of limited human and financial resources. In addition, drought, locust swarms and rinderpest devastated farms during the 1890s and nearly suspended the annual Show on at least one occasion.

Despite all these setbacks there were several developments that promoted a central annual Show in Pietermaritzburg. From 1857 horse breeding made its appearance there with some farmers, led by three-term president Charles Barter of the Karkloof, having imported good-quality bloodstock to exhibit along with their progeny. They were not only intended for local hunting and racing but also for export to Mauritius and India. From 1861 the Society was able to use the city's first market house, for a monthly rental of £2, where the Produce Section of the Show was exhibited. Two years later the initial annual state grant-in-aid was doubled to £100.

The 1867 Show featured a new section for Colonial Industry with a £1 prize from 1868 that attracted some unusual items, including innovative agricultural inventions. The 1867 Show also witnessed the first competition for winter cattle fodder, the shortage of which during the region's dry mid-year months was often aggravated by veld fires. That year a Farmers' Club was established in Pietermaritzburg to provide a discussion venue and reading room with relevant agricultural literature. Following the club's demise in 1874 George Sutton continued to supply farmers with useful information in his *Natal Witness* column under the pseudonym Agricola.

The introduction in 1869 of a jumping competition in imitation of the English Royal Agricultural Society met with a mixed response as some considered it inappropriate for an agricultural show. It was subsequently excluded from the programme with no anticipation of the important role it was to play in future years.

By the 1870s Pietermaritzburg's May Week was attracting attention from all over the Colony with the Agricultural Society's Show rivaling a birthday ball in honour of Queen Victoria and Fort Napier's annual military review in importance. Other events that week included horse races, amateur dramatics and church bazaars in addition to the Show dinner and numerous privately organised social gatherings. Perhaps for these reasons the Show was becoming a sartorially more elegant event than ever before.

The arrival in Pietermaritzburg of the railhead from Durban in 1880, followed by its advance to the Transvaal border in 1891 and the subsequent construction of branch lines into the countryside, further helped to promote the Show by facilitating the movement of exhibits and visitors.

The acquisition of a permanent showground provided an even bigger boost to its fortunes. It was first proposed in the mid-1870s as the number of exhibitors and attendees increased. However, this was delayed by lack of funds and by the vigorous opposition of businessmen who feared that their premises surrounding the Market Square show site would suffer a serious loss of sales.

In 1889, with the support of the City Council, the Show at last found a 3.2 hectare home of its own on what is now the Drill Hall site behind the old cemetery and still conveniently close to the city centre. This was developed with a generous £1 500 loan from Henry Fell, a successful Eston farmer.

It now became possible to accommodate livestock from distant farms both before and after judging as well as comfortably displaying a greater variety of other exhibits.

For the first time the 1889 Show was prefaced by an official opening, on this occasion featuring the British Governor Sir Arthur Havelock. In 1891 the Pietermaritzburg Agricultural Society was instrumental in forming the Natal Farmers' Conference to improve closer collaboration within the Colony's farming community. The Conference became the Natal Agricultural Union in 1905 and by 1924 it had strengthened from an initial nine to 75 associated societies.

Meanwhile further improvements were effected to the new Showground. In 1891, when electric lighting and a telephone were acquired, the event was extended to two days. In 1893 it attracted a record 1 095 exhibit entries and nearly 6 000 visitors, some from beyond the Drakensberg. That year the Society's first professional secretary, A. Whittle Herbert (1892–1903), successfully made the Show a special occasion to celebrate the Colony's acquisition of responsible government as well as the 50th anniversary of formal British annexation.

In 1896 an additional 1.6 hectares could be afforded by letting the premises out for other events during the course of each year. However, before the turn of the century it was again increasingly evident that the Show needed a bigger site, preferably close to the railway line for the convenience of livestock and other exhibitors.

In 1900, following the outbreak of the Anglo-Boer War (1899–1902), the Drill Hall property was commandeered for military purposes and no shows were held for two years. During that time the Agricultural Society's activities were moved to a leased site of 8.4 hectares situated south-west of the Dorpspruit between Boshoff Street and what was then Commercial Road (Outspan 4). It was no further from the city centre than its predecessor and adjacent to the Greytown railway line.

The government's purchase of the Drill Hall site for £9 000 enabled the Society to pay off the £2 000 it still owed on the bond and to spend the balance developing its new permanent premises. The municipality agreed to hold the property in trust for the Society provided at least three city councillors sat on its committee and it was used only for agricultural exhibitions except with its special approval.

Arrangements for the 1902 Show (the first since 1852 to be held in June) were still fairly makeshift, with officialdom accommodated in a tent. By the following year, among other amenities, there was a permanent produce hall and a grandstand made with locally produced bricks.

Other significant developments during the early years of the new century included the introduction of electric lighting into several buildings, the extension of the municipal tramline from the city centre to the Showground gate on Commercial Road and the installation of turnstiles and a bandstand. The establishment in 1902 of the Cedara Agricultural Research Station, where an agricultural college emerged, marked the beginning of a relationship with the Society that soon proved to be mutually beneficial; as was the link with Weston Agricultural College in Mooi River from 1914.

A change of name for the Society was considered, off and on, for some time. Eventually, in 1904, it was suggested to Natal's Prime Minister Sir George Sutton that the Show's prizes would be greatly enhanced if 'presented by a body having Royal sanction'. This would make them quite distinct from the region's other shows. The proposal won official support. In November that year the Pietermaritzburg Agricultural Society was informed that the recently crowned King Edward VII had given permission for it to be known as the Royal Agricultural Society of Natal. A new era in its history had begun.[2]

The Royal Agricultural Society (1904–1983), or RAS as it came to be known, experienced a mixture of setbacks and advances much as its predecessor had done. Unsurprisingly, natural disaster played its part. East coast fever so ravaged the region that the Cattle Section was closed between 1907 and 1910 while the disease was brought under control. A drought further compounded the misfortune of farmers and contributed to a decline in the Society's membership from 464 to 210 between 1904 and 1908.

Government's temporary withdrawal of its annual grant, now £500, deepened the financial crisis. Despite various economies, including the substitution of gold with silver trophies, for a time the Society had to survive on an overdraft. This dire situation was relieved by the £2 000 it received as its share when in 1910 Natal's government distributed a cash surplus on joining the new Union of South Africa. Moreover, not only was the annual state grant renewed but the Union government undertook to assist the RAS in making permanent improvements to its premises on a £-for-£ basis.

In 1933 there was another drought and in 1947 the Dorpspruit flooded, resulting in extensive damage to several buildings. Further flooding delayed the opening of the 1965 Show and nearly led to its cancellation. However, for the most part, the Society's misfortunes were man-made. The 1906 Show was cancelled due to the Bhambatha uprising against the poll tax. This disrupted farming activities in several areas and spread the now familiar scourge of East coast fever through the uncontrolled movement of cattle.

The Show survived the Great Depression of the early 1930s but suffered declining income with fewer exhibits in the Industrial Section and lower attendance figures until 1935. While buildings were leased out for other purposes to relieve the financial situation, it was the annual provincial and municipal grants of £500 and £350 respectively that saved the day. This helped to compensate for yet another suspension of the central government's grant.

The economic disruption and loss of manpower during the First World War (1914–1918) had minimal impact on the RAS compared with that of the Second (1939–1945). The 1940 Show took place despite a 25% decline in entrants but none were held between 1941 and 1946. The Showgrounds and buildings were leased to the Department of Defence at £300 a year, but the resultant wear and tear to its facilities subsequently had to be repaired.

Despite recurring crises, financial and otherwise, the Society continued to introduce significant innovations and to improve its showground facilities. In 1904 it was instrumental in forming the Natal Agricultural Judges' Association with other societies in the region. This served to standardise judging criteria and to compile a register of competent individuals from which member societies could benefit.

The admission of women as RAS members in 1909 at 50% subscription rates was seen as a means of boosting its flagging finances. They had long been involved in the Shows as exhibitors under their husband's names and as voluntary assistants. They were soon running the tearooms and in 1914 were allocated much more space in the main hall to accommodate their increasing exhibits. Although not initially permitted to take charge of their own Crafts Section they soon gained representation on the executive committee.

By 1920 there were more than 400 female members and in 1923 the Crafts and Industries Section acquired autonomy with its own committee. Two years later, having previously met in their own homes, the women were given office space at the Showgrounds. By 1924 they had raised sufficient funds to build their own hall for what became known as the Crafts and Home Industries Section.

Yet another important development was the acquisition of more land. The 3.2 hectares across Commercial Road which had recently been leased from the Railways Department at £1 a year was not retained and from 1913 became a car park. However, the outright purchase of land north-east of the Dorpspruit for £650 from a deceased estate was significant for future expansion as the

Showground became the Showgrounds.

The existing house there was subsequently renovated for the caretaker's use, cattle sheds were built, trees planted to secure the stream's banks and tenders called for the construction of a suitable bridge. Back across the river in 1912 a new grandstand was constructed complete with a judges' box and motor cars were displayed for the first time. The growing importance of industrial exhibits was demonstrated when they were accommodated in the brick hall originally intended for the Produce Section.

After leasing the Showgrounds to government in 1916 as a wartime remount depot the Society paid for all the improvements the military had effected, including electric lighting. In addition, a new ring and stand were installed for stock sales and internal roads were hardened. After the war, further leased land was acquired beyond the Dorpspruit towards Hyslop Road and the municipality granted the RAS an additional plot with another 3.6 hectares eventually being added to the Cattle Section in 1927.

There were noticeable improvements in the quality of animals exhibited in the 1920s as well as a greater variety of arena events. The 1935 Silver Jubilee Show in honour of King George V's 25 years on the throne, featured a searchlight tattoo by the special service battalion from Pretoria for two consecutive nights. The Olympia Motor Hall was constructed during the 1930s and connected to the Main Arena by a new bridge across the Dorpspruit. In 1939 the expanding Crafts and Home Industries Section was moved into that hall, but initially had to contend with a rough gravel floor and leaking roof.

By 1953, when the Society resolved to celebrate its centenary, the Showgrounds had been developed almost beyond recognition with more land acquired beyond the Dorpspruit and numerous facilities added. The more than 3 000 RAS members and the general public were treated to a lavish Centenary Show while downtown the occasion was recognised with a City Hall ball and a performance of *H.M.S. Pinafore* at the Rowe Theatre.

The ongoing success of the RAS and its annual Show had relied heavily upon public support, not least from the agricultural and commercial-industrial sectors, on far-seeing leadership and on sound administration, especially in times of financial hardship. It was secretary/manager A.C. (Bertie) Bircher (1940, 1945–1973) who helped to organise the year-to-year lease of the members' pavilion, wooden grandstand and arena to the newly formed Collegians Club (1950) for their rugby and cricket matches during all but two weeks each year while changes were being undertaken at the Woodburn grounds. This arrangement provided the Society with another welcome if temporary additional source of regular income.[3]

After the 1953 centenary celebrations similar arrangements were made with other organisations for varying lengths of time as a means of ensuring that the Society stayed afloat. New arena attractions intended to improve gate takings at the Show included performances by the Dagenham Girl Pipers, Major George Iwanowski's Lipizzaner horses and the internationally recognised flag throwers of Arezzo.

The more expensive of these events required substantial sponsorships which also became an increasingly important feature of other sections of each Show. By the mid-1950s there were as many as 25 section sub-committees reporting to the general committee which provides an indication of the ongoing growth of the RAS and its Show.

Despite financial constraints there were further improvements and additions to the facilities. The new grandstand was finished in 1954, complete with dining room, kitchen and upstairs offices. A new industrial hall was added and the wood and iron stables were removed from the vicinity of the Main Arena. These were replaced with 100 new boxes near Chatterton Road, off which a new main entrance was constructed. New bridges were built across the Dorpspruit, including one for the exclusive use of horses and new accommodation provided for farmworkers taking care of animals during shows.

After South Africa became a Republic in 1961 the suggestion that the Society should again change its name was firmly rejected. It remained the Royal Agricultural Society with permission successfully sought from the Administrator to use the Natal crest until it acquired its own coat of arms. The 1965 proposal that the RAS should again move its Showgrounds to a larger site, this time to Mkondeni, was also not pursued. This was partly for sentimental reasons, but also out of concern for the commensurate loss of its now considerable facilities and the prospect of assuming an enormous financial debt it would struggle to overcome.

Instead, it was decided to re-plan and improve the existing premises. Floodlighting was installed for the April 1966 Show so that exhibits could be viewed at night and the following year it reverted to the month of May to avoid clashing with the Rand Show. In 1974 the University of Natal's Professor of Architecture Paul Connell was commissioned to assist in developing a master plan, which provided useful guidelines for subsequent development.

In 1976 what soon proved to be an important new attraction was introduced in the form of the September Garden Show, which the Natal Parks Board and Orchid Society, among others, enthusiastically supported.

During the 1970s catering for members was entrusted to the Victoria Club while various food stalls continued to raise funds for worthy causes. The funfair was improved, better seating was installed in the Main Arena opposite the grandstand as well as new accommodation for Barclays Bank. In 1977 the first service organisation in the form of the Duzi Steak Inn erected its own building on the premises. The following year the Tongaat farm area was established followed in 1979 by Standard Bank's exhibition building and in 1980 by the City of Pietermaritzburg Hall.

Stock Owners, whose stud sales were always a highlight of the Show, financed new cattle and pig sales rings in 1975 and 1978 as well as converting their pavilion into a double-storey building in 1983. By then the existing Angus, Bonsmara, Brahman, Jersey and Simmentaler clubhouses had been joined by similar structures for the Charolais, Friesland and Hereford breed societies. The Meat Board, Stock Owners and Standard Bank, among others, all contributed generous livestock prizes. Yearling sales were also an important event with Snowdonia selling for a record R240 000 in 1983.

Equestrian events, considered inappropriate in the 1870s, now attracted considerable public attention. There were as many as 20 000 Main Arena spectators, with provincial and national championships sometimes being held during the Show. Attendance figures continued to improve despite rising inflation-driven admission charges and increased stall rentals.

In 1976 there was concern about overcrowding following the sudden large influx of schoolchildren to the Show. Agreement was subsequently reached with all the ethnically separated education authorities that admission would be limited to standards eight to ten (the last three years of secondary schooling).

During the 1940s consideration had been given to admitting black farmers as exhibitors. This proposal was rejected in favour of assisting them to arrange their own shows and subsequently to hold one in Pietermaritzburg after the annual RAS event. Eventually, in May 1978, a permit was secured for a multiracial Show that included blacks as exhibitors and attendees. The benefit of their involvement in the former capacity was immediately felt, particularly in the Crafts and Home Industries Section.

In 1981 attendance declined from the previous year's 158 709 to 132 540 when black communities boycotted what for that year was contentiously named the Republic Festival Show. This was in recognition of the twentieth anniversary of South Africa's republican status and in exchange for R100 000 from government with which to erect a new grandstand.

By 1983 attendance had risen again to 191 926, still far short of the record 217 305 in 1977 but despite unfavourable economic conditions and the worst regional drought in living memory. This led to the closure of the already declining Produce Section which had once been such a prominent feature of earlier shows, but fortunately other sections continued to thrive.

In 1974 the RAS office staff moved from its tiny first-floor premises at 18 Timber Street to new accommodation at the Showgrounds inside the members' grandstand. That year Mark Gaye Shute, who had served the Rand Show in a similar capacity, succeeded Bertie Bircher as general manager. He was immediately concerned with preparations for the Society's 125th Anniversary Show in 1976, which introduced its recently registered coat of arms. This symbolically depicted the influences that formed part of its history and incorporated the Latin motto *Agri Cultura Robur Rei Publicae* (Agriculture is the Strength of the State).

Shute's arrival led to a number of administrative and organisational improvements. In part these were influenced by a visit to several shows in Britain during 1978, but he also had firm foundations on which to build. They had been provided by those who in previous decades contributed in many different ways to the RAS and its annual Royal Show. Apart from the support of the general public there had been numerous generous corporate and private donors, as well as countless regular exhibitors, competent auctioneers, conscientious judges, reliable stewards and enthusiastic equestrian clerks of the ring, not to mention all the women who had voluntarily worked in the Crafts and Home Industries Section and in other capacities.

Ashton Tarr had served for years as honorary veterinary surgeon, T.K. Allison had maintained firm control of gate finances and Max Doepking was a similarly efficient groundsman (1946–1974). From 1953 Major Percy Lewis had been the Show's chief announcer, followed by the soon similarly familiar voices of Robin Alexander and Dave Walmsley. There had been many dependable office staffers like honorary life member Pauline Peacock. The RAS and its predecessor had also enjoyed a succession of efficient secretaries/ managers as well as the leadership of long-serving committee members and presidents.

Such loyalty and dedication in so many different categories of endeavour would continue to be vital in the future.[4]

ENDNOTES

1   For further details of the colonial setting in which the Pietermaritzburg Agricultural Society and its successor the RAS was established see: Bill Guest and John M. Sellers (eds), *Enterprise and Exploitation in a Victorian Colony* (Pietermaritzburg: University of Natal Press, 1985) : 8–10, 13; John Laband and Robert Haswell (eds), *Pietermaritzburg 1838–1988: A New Portrait of an African City* (Pietermaritzburg: University of Natal Press and Shuter & Shooter, 1988): 24–32, 120–125; Andrew Duminy and Bill Guest (eds), *Natal and Zululand from Earliest Times to 1910* (Pietermaritzburg: University of Natal Press and Shuter & Shooter, 1989): 116–132, 302–305; Bill Guest, *A Century of Science and Service: The Natal Museum in a Changing South Africa, 1904–2004* (Pietermaritzburg: Natal Museum, 2006): 1–10.

2   For a more detailed account of the Pietermaritzburg Agricultural Society see: Lindsay Young, *A History of the Royal Agricultural Society of Natal, 1851–1953* (Pietermaritzburg: RAS, 1953): 7–47; Ruth Gordon, *Natal's Royal Show* (Pietermaritzburg: Shuter & Shooter, 1984): 8–17, 23–47.

3   For a more detailed account of the RAS between 1904 and 1953 see: Young, *A History of the Royal Agricultural Society of Natal, 1851–1953*: 47–65; Gordon, *Natal's Royal Show*: 47–77.

4   For a more detailed account of the RAS between 1953 and 1983 see: Gordon, *Natal's Royal Show*: 78–90, 92, 99–103, 105, 108–119, 121–125, 130, 131, 134.

1 AN ERA OF 'GOOD JUDGMENT AND A FIRM HAND', 1984–1994

THE LAST DECADE of what Ruth Gordon described as 'the era of Mark Shute',[1] ending with his retirement in 1994, witnessed among other developments numerous novel Main Arena and additional attractions at the annual Royal Show. There were also new records and achievements in the various livestock and other sections, further improvements to the Showgrounds facilities and fluctuating attendance figures.

The **125th Royal Show**, held from 25 May to 2 June 1984, was considered at the time to be 'the best supported, the best attended and the best presented Show in the history of the Society'.[2] The impressive attendance figure of 207 341 was boosted by the unusual inclusion of two public holidays (Republic Day and Ascension Day) but failed to exceed the all-time 1977 record of 217 305. This had been achieved with the long-remembered assistance of school groups which had contributed 45 625 individuals through the turnstiles compared with only 11 560 in 1984.

RAS President R. (Ron) McDonald nevertheless justifiably described the 125th as 'a momentous occasion', held under clear blue skies and with the dreadful 1983 drought safely broken. An able managerial team, including vice-presidents John Fowler and Jeremy Snaith, general manager Mark Shute and the various section committees, also helped to ensure that it was indeed a great success.

The Pietermaritzburg Turf Club opened the 1984 Show season when on Saturday 14 April it hosted a race meeting for the RAS. Champion jockey Michael Roberts steered Rain Forest to victory in the 1 400 metre 125th Royal Show Handicap while the Society's members snapped up special souvenirs in the form of ashtrays, cuff links, scarves and ties. The *Natal Witness* produced a facsimile of its 1851 edition which reported the origins of the RAS and Ruth Gordon's *Natal's Royal Show* was published to update its subsequent history.

The Society's image was greatly enhanced by the creative artwork with which Clive Hatton Design Studios had advertised each Show over many years. The 125th's logo was prominent for several months in the media and

Ron McDonald's involvement with the Show went back to the days when all the Showgrounds roads were surfaced with wattle bark, and livestock was delivered at the neighbouring Victoria railway siding from which plumes of smoke wafted across the Main Arena. Still in his early twenties, during Show time he and his 'wonderfully capable' assistant Elliot Mbembe regularly slept overnight in a shed which the family firm McDonald's Seeds and Feeds (1902) had erected in the Cattle Section. From there they supplied livestock feed, including lucerne, hay and cubes at all hours to needy exhibitors.

He recalled nights spent at the Showgrounds that were so cold by the early mornings the water jump in the Main Arena was frozen solid enough to provide a skating rink for the resident geese! Indeed, the weather was so severe that Garth Carpenter asked permission to accommodate his nearby exhibition of snakes overnight in the relative warmth of McDonald's premises. Fortunately, only one ever escaped from its paper bag, never to be seen again!

Ron McDonald subsequently became involved in the then large Produce Section and eventually served as one of its judges, as well as in the RAS committee structure culminating in his six-year term as president (1981–1987).[3]

on award cards, commemorative beer cans, decorations, envelope stickers and special medallions. The RAS also presented long service awards to all stockmen who had attended at least ten previous Shows. Guests of honour Justice and Mrs John Milne presented 22 of them during the official opening programme.[4]

Administrative and grounds staff made their usual unobtrusive yet vital contribution to the success of the Show by adequately preparing all the facilities and ensuring their smooth functioning. This was particularly true of grounds manager Don Byres and of Bert Cornell, Ken Easthorpe and Danie van Wyk with regard to the public address system, electrical installations and gate control respectively. They were by then all experienced members of what Ron McDonald called 'the A team' who maintained the Showgrounds throughout the year.

Unfortunately, from time to time, there were incidents that were entirely beyond the A team's control. The **Main Arena displays** in 1984 were marred by a fatal accident involving the renowned Mexican highwire artist Chuchin. He was what Ron McDonald described as 'an amazing little gentleman' who

Bert Cornell was only 17 years old when in 1957 he arrived in Pietermaritzburg. He first became acquainted with the Showgrounds the following year on being sent to undertake tasks there while serving his apprenticeship with Reid's Radio. He was running his own business, Radio Cornell, when in 1974 the new RAS general manager Mark Shute invited him to take charge of the Society's public address system, which was very old and dated back to the pre-1939 era.

On Shute's insistence Cornell's first task as part of a general sprucing up campaign was to install underground cabling in place of the tatty overhead wiring which sagged along a series of wattle poles in the Main Arena. He was also required to install 40 second-hand loudspeakers which Shute had brought with him from the Rand Show. What was initially merely a small part of Bert Cornell's contract business became increasingly time-consuming over the years and a significant dimension of his life.[5]

prior to the tragedy had been entertaining children by tying balloons into the shapes of animals. During the Show he had already endeared himself to the public but, while descending at the end of his seventh performance in unfavourable, drizzly conditions, his safety rope slipped from the hook of the crane and he fell to his death in front of a horrified crowd.

While it was subsequently decided that in future all aerial acts at the Show should require a compulsory safety net the RAS was left with the unhappy task of tracing Chuchin's family and repatriating his body. A time of celebration suddenly became one of the darkest moments in the history of the Society. The tragedy cast a gloom over the Main Arena which was only gradually lifted by the other displays that year.[6]

They included those of the South African National Defence Force (SANDF) Equestrian Centre's motorcycle team, the Pietermaritzburg high schools' massed gymnastic group, the South African Police (SAP) Wentworth Training College physical training squad, the University of Natal's drum majorettes, the T-H security dog school and the Dogmor dog-jumping competition.

Additional attractions also contributed to the success of the 125th Royal Show, including the funfair's looping star and the popular rotary roundabout transportation facility. Equestrian events were as popular as ever.

The **Horse Section** indeed attracted the most attention with its show jumping competitions. International course designer Harold Preston and the trilingual contribution of entertaining commentator Robin Alexander helped, as before, to ensure their success. As at other Shows whenever the hooter of a passing train could be heard from the nearby railway line Alexander would declare 'Van Blerk [the anonymous driver] is at it again!'

Bert Cornell remembered regularly taking delight in telling Preston that what he regarded as 'a challenging course' was actually 'diabolical'. Their friendship was based on a common love of the Royal Show and of opera, although years later a disagreement about the volume level of the public address system in the Main Arena nearly led to Cornell's angry resignation![7]

In 1984 the prestigious show jumping championship of South Africa, coupled with R33 000 in prize money and the presence of British judge Harry Hindle, ensured the participation of the country's leading riders and horses. All classes were filled, including a record entry of 92 A-grade show jumpers.

No less than six other record entries in the competitive sections provided a further important indication of the success of the 125th Royal Show. They included 1 757 cattle, 601 sheep, 348 pigs, an overall 7 167 livestock entries (including other categories) and a grand total of 13 469 in all sections with a record 6 187 in Crafts and Home Industries.

The **Cattle Section**, like the Horse Section, enjoyed a boost in entries following the decision to hold the national championships of the Angus and Sussex Breed societies at the 125th Show. Glen Klippenstein from Missouri (USA) and Hamish Smith from Zimbabwe judged in these two categories respectively.

In what section committee chairman George Poole described as 'the best-ever show of cattle at the Royal' ten beef breeds were exhibited, Aberdeen Angus being the best supported with 168 entries and thirteen exhibitors followed by Sussex with 157 and eleven. John Fowler, RAS vice-president (1984–1987), celebrated the occasion by displaying the number plate 'Angus 1' on his Rolls Royce. This was by kind permission of Pietermaritzburg's mayor Pam Reid who was a great supporter of the Royal Show and recognised the benefits, financial and otherwise, it brought to the city.

Jerseys were the strongest dairy breed on show with 157 entries and eight exhibitors while the 306 Simmentalers and their fifteen exhibitors dominated the dual-purpose category. The welcome number of entries did cause 'cramped conditions' which pointed to the need for improved facilities and to the early release of dairy breeds from the Show due to the loss of milk production.[8]

The pedigree cattle sale that year saw J's Harlequin Stud near Kokstad achieving the top price of R11 750 for a Simmentaler bull with 143 of the nearly 300 head entered selling for a R329 500 aggregate. This was a 37% increase on 1983 and amounted to an overall average of R2 304, which was only R345 lower than the 1982 record. The two breeds most in demand were Simmentaler with 40 sold at an average R2 485 and Sussex with 30 sold at an average R2 982. Mapstone Bros of Baynesfield won both top awards in the national carcass competition for the second year running.

The **Sheep Section** also enjoyed a successful Show, holding two national championships with Hampshire and Suffolk breeds attracting 163 entries from seven exhibitors and 126 from six respectively. There were also an impressive 152 Ile de France entrants from six exhibitors with demand for them being strongest at the pedigree sheep sale. This event amounted to a record R16 400 compared with R10 605 the previous year and achieved a record average price of R586 compared with R530, although only a modest 28 of the 75 head offered were sold.[9]

The **Pig Section** similarly achieved its record entry with a national championship to celebrate the 125th Royal Show. This created a livestock accommodation problem giving rise to the clear need for a reduction in the number of animals on show. As in the Cattle Section, voluntary reductions on the part of exhibitors overcame the crisis.[10]

Landrace were predominant with 116 entries and nine exhibitors, followed by Large White, Duroc and Hampshire. The aggregate sale of 53 pedigree pigs for R37 850 and the average price of R714 were only slightly lower than the records achieved in 1983. However, J.H. Meschede of Bryanston achieved a Show record R3 600 sale for a Large White boar.

The significant contributions made towards the 125th were not confined to the major livestock categories. The **Apiarian (Honey), Dairy Produce, Budgerigar and Cage Bird, Poultry, Pigeon and Rabbit** sections all added, as in previous years, to the Show's educational and general interest value. While the Apiarian Section was feeling the need for a new building the 49 entries in the Dairy Produce Section were higher than the previous year but considered 'disappointing' and a 'cause for concern'.[11]

The **Crafts and Home Industries Section** attracted a record 6 187 entries. These comprised 2 761 from 548 exhibitors in the senior category, 407 from 133 juniors, 1 791 entries from schools, 1 065 from women's institutes and 163 participants in the Norton knitting championships.

The **Commerce and Industries Section** aroused much public interest with its Royal Motor Show mounted by the National Association of Automobile Manufacturers of South Africa. The association's eleven members exhibited a variety of special imports and flagship models in an Amplihall structure covering 660m² of the picnic area at the northern end of the Crafts Hall. This innovation proved significantly cost-effective for exhibitors by excluding the need for high-cost display material.

Among the hall exhibits those of the City of Pietermaritzburg and of the Department of Posts were of an exceptionally high standard. However, hall exhibitors won only nine of the 35 medallions awarded for displays in the Commerce and Industries Section with twelve going to those on open sites and fourteen to those in individual pavilions.

Physical improvements, timeously completed, were also essential to the success of the 1984 Show. These involved an ambitious programme of unprecedented expenditure amounting to R550 000. It included a new pedestrian entrance on Commercial Road, upgrading the members' clubhouse, and modernising the dining room and bar.

It also involved re-roofing the Crafts Hall, a new milk bar in Block E, an additional block of horse stables, new cattle stall block and office for the chief cattle steward, roadway paving around the cattle judging ring and additions to the Sheep Section's accommodation with new pens and paved walkways. The ash surface, which had previously covered all the roads and pathways, gradually began to disappear. The Society's dedicated 125th special committee under J.S. (Jeremy) Snaith's leadership generated donations exceeding R330 000 to finance these improvements.

Mayor Pamela Reid's Pietermaritzburg City Council made the largest contribution for the Commercial Road public entrance built over the Dorpspruit. The three dozen other donors included First National Bank (FNB) (for the dining room and bar), Hulett Aluminium (Crafts Hall re-roofing), the Dairy Board and National Co-operative Dairies (NCD)-Clover (the new milk bar), the Natal Provincial Administration (NPA), Stock Owners Co-operative and the Tongaat-Hulett Group (various improvements to the Cattle Section) and Wolhuter Steel (upgrading the Sheep Section).[12] Other upgrades and new buildings were also evident with Geisers, H.M. Leers, the Natal Angus Club, Santa Gertrudis Club and Taurus Co-operative leading the way.

The 125th Royal Show indeed proved in many respects to be a 'momentous occasion', even though it was only one of fourteen shows in the subcontinent at the time, including that of Zimbabwe.

Attendance at the 1985 Show declined, more spectacularly than anticipated, from 207 341 at the 125th to 166 509. The prevailing economic recession and a 50-cent increase to the R2.50 adult gate admission charges (R1.50 for evening sessions) had an adverse impact. This was unexpectedly compounded by cold, rainy weather and extensive N3 road works at Key Ridge that created difficult traffic conditions between Durban and Pietermaritzburg.[13]

As always, the vagaries of the weather and the inclusion/exclusion of public holidays during the show period affected attendance figures. These recovered to a new record 221 974 in 1988, even though school pupil attendance numbered less than 9 000. The total slumped again to 206 615 in 1990 although this was still 14 000 higher than the average annual attendance for the decade.

In 1986, with a national state of emergency seemingly imminent, the first nine-day Royal Show was held; the first show in South Africa to open on a Sunday, with an encouraging 16 244 attendance on that day. This followed recent bloodstock sales, a book fair and garden show that had all opened on Sundays. An evening session attendance record of 7 662 was achieved by the popular fireworks night, rising to 13 692 in 1988.[14]

A record total attendance of 34 179 for all the evening sessions in 1986, rising to 43 494 in 1988, was partly attained by implementing a 5.30 pm gate price reduction. In 1987 a record 26 818 attendance on the last Saturday of the Show was registered. This was followed by a record Sunday attendance of 17 072 in 1989, rising to 19 800 in 1991 when a record attendance for the sixth day (Wednesday) of 33 762 was also achieved.

There was some concern about inadequate advertising of the Royal Show in the Durban area and about the fact that it primarily attracted members of the white and Indian communities with more effort needed to encourage black visitors and exhibitors. Indeed, the black community and greater Durban were perceived as 'the two major growth areas' as far as increasing attendance was concerned.

For the first time in six years in 1993 the Show did not attract an attendance in excess of 200 000 (only 169 647) despite fine weather and the extension of its duration from nine to ten days. These factors failed to compensate for the mounting political tension countrywide in the build-up to South Africa's first democratic elections.

South African Communist Party (SACP) leader Chris Hani had been assassinated on 10 April just five weeks before the 1993 Show and, amid fears of possible civil war or of disturbances within the Showgrounds, security staff at the main entrance were placed on a round-the-clock alert. The situation was not improved by a recurrence of depressed economic conditions, slight increases in admission charges the previous year and a teachers' strike which substantially reduced the traditional school group visits.

The following year the Show ran again for ten days and attendance improved to 183 624 with the assistance of a generally more relaxed atmosphere just a week after South Africa's first multiracial democratic election. Indeed, the Show's slogan that year was 'Relax at the Royal – This one's for you!'

There was also an additional public holiday (10 May) when Nelson Rolihlahla Mandela was inaugurated as the new State President but it was to be the last Ascension Day public holiday. Appropriately, John Hall, chair of the National Peace Committee, opened what was the 135th Show. He declared it to be 'a day for great optimism being the dawn of the new fully democratic South Africa with increased hope for the future'.[15]

An analysis of the last two decades established that average attendance for the ten years ending in 1994 had exceeded that of the previous ten years by 10% whereas the average daily attendance reflected a decrease in recent years due to the lengthening of the Show to ten days. The proportion of paid adults to total attendance in 1994 was exactly the same as the average percentage for the last seventeen years and the only record achieved in 1994 was the 19 803 who attended the Show on the last Sunday.

In his report to the management committee in June 1993 the president Ron Glaister had already observed: 'I feel that the Show has reached a crossroads in regard to both content and direction. The future success of the Show will increasingly rely on the support of the black community'.[16]

Main Arena displays still included an official opening held on the first Wednesday of each Show. The guest(s) of honour, executive members and spouses visited each exhibition hall in accordance with the grounds manager's carefully scheduled programme followed by a formal lunch. Then they proceeded to the members' grandstand to witness the grand parade of prize-winning livestock and presentation of trophies, after which the Natal Carbineers' Band usually concluded the occasion with a rendition of the 'Last Post'.[18]

Among other Main Arena events during the 1980s and 1990s were perennial favourites such as the beating of the retreat ceremony, which the Carbineers and (in 1992) 121 Battalion performed. There were also Scully Levin's and the Winfield Magnum Team's pitts special aerobatics, skydiving, equestrian, gymnastic, drum majorette, dog jumping and sheepdog exhibitions as well as the fireworks nights and concluding light classical, country and western, folk and pop concerts. Several innovations included the riveting Simba wheel of death or cloud swing, rifle drills, a lumberjack competition, vintage sports car, steam engine and tractor parades, in addition to motor and motorcycle stunts.

R.J. (Ron) Glaister was a Pietermaritzburg businessman who was very active in the life of the city, including service on its Council as its youngest member. The RAS was one of several interests and during the course of many years he contributed to its activities in various capacities.

These included vice-president (1988–1990), president (1991–1995) and chair of the capital projects committee (1996–2016). He was very involved in the maintenance and physical improvement of the Showgrounds, including the design of some of the new structures that were erected there. Glaister had a particular interest in the Commerce and Industries Section but was enthusiastic about every aspect of the annual Show which, like so many others, became an important part of his life.

Mrs Paula Glaister, who actively supported her husband's involvement and worked in the Duzi Steak Inn, was awarded honorary life membership when he became an honorary life president. The presidential gavel that Glaister presented to the executive ten years earlier was a reminder both of his service to the RAS and of his woodworking skill.[17]

In 1987 the SADF contributed prominently to the Main Arena programme in celebration of its 75th anniversary, with Chief of the Army Lieutenant-General and Mrs A.J. Liebenberg as guests of honour. The Defence Force's participation resulted in a considerable reduction in Main Arena costs that year. However, there was some criticism of the Society for providing a venue for this celebration at a time of international sanctions against South Africa and its military involvement in neighbouring countries.[19]

The 1988 Show was linked to the City of Pietermaritzburg's 150th anniversary celebrations for which its Council decided upon a low-key approach with 'no major festivities', apart from emphasising the local 'annual attractions' including the Royal Show. RAS president John Fowler nevertheless declared it to be 'in every respect, the most successful Royal Show ever' as well as 'the leading public attraction' in that celebratory year.[20]

There were several innovations that helped to achieve the record overall 221 974 through the turnstiles, including a 50% increase in Main Arena attendance compared with the previous year. The Police Force was celebrating its own 75th anniversary and provided the core of the arena programme, as the Defence Force had done the year before, with six different displays in addition to two exhibits in the Commerce and Industries Section. In the wake of the previous year's experience an effort was made to 'include sufficient civilian displays to dilute the impact of the SAP presence'.[21]

Supporting arena performances included massed aerobics, Portuguese folk dancing, the National Volkspele Group and the Maritzburg College band as well as a strong man challenge. The parade of commercial cattle was another successful innovation which appropriately helped to underline the agricultural origins and emphasis of the Show, soon becoming a regular opening day event.

Overhead the 'masters of the sky' aerial trapeze artists and police helicopters provided considerable excitement, as did the formation flying of the South African Air Force's (SAAF) Silver Falcons, which was a first not only for the Show but also for the city. The evening sessions were greatly enhanced by the installation in 1986 of new floodlighting equipment which virtually eliminated all shadows and quadrupled the intensity of the old system that had been in operation for twenty years.

By the early 1990s it was felt that there was a 'degree of sameness' about the Main Arena displays and a need 'to overcome the public's view that the Show was little changed from year to year'. It was pointed out that the Rand Show now featured several small entertainment areas catering for a variety of tastes instead of concentrating all special events in the Main Arena. The arena directors' committee was finding it difficult to find suitable acts, particularly those that appealed to the young. This was aggravated by the fact that the Defence Force, Police and Correctional Services were offering fewer options than in previous years.[22]

New attractions were nevertheless unearthed, including the Amabele Breweries tug-of-war competition, outstanding massed displays by the SANDF Women's College from George, Chubb aerobatics, Natal Field Artillery (NFA) gun demonstrations, firefighting displays involving an enormous Russian helicopter, Pony Club mounted games, Belgotex night polo, Jack Russell races (including unrehearsed squabbles en route), the marching pipe bands of Caledonian societies and bungee jumping.

Seemingly, the latter did not attract any misguided participants from the nearby Foaming Tankard beer garden! Ken Easthorpe recalled that intoxicated students were certainly drawn to the swimming pools on display at each Show and often warmed themselves afterwards at the heated police station near the main gate.[23]

Additional attractions outside the Main Arena were the 1985 Hall of Kitchens, a group exhibit that nine suppliers of kitchen fittings set up in the Commerce and Industries Section which drew much public attention. So too did the NPA Theatre the following year when it was transferred to the Crafts Hall from the Durban Expo with its sophisticated audiovisual production '76 Years of Responsible Government in Natal'.

In 1986 Dhoda's Market, the Hilton Round Table and Scott's Caterers, all loyal supporters of the Show, erected new food outlets that improved the facilities available to the public while the Imperial Hotel provided a catering service to RAS members.[24]

The following year Natal Command and Natal Medical Command staged static exhibits and there was also a display of military art as part of the Defence Force's 75th anniversary. In 1988 consideration was given to following the example of the Rand Show in giving less emphasis to Main Arena attractions and more to pop music at some other venue. On reflection, it was felt that this would be out of character with the atmosphere of the Royal Show.

The funfair continued to be as popular as ever and Ron McDonald recalled how as president he and general manager Mark Shute were also able to enjoy it during show time, in their own way. Every evening after closure they concluded their inspection of the grounds with relaxing nightcaps in the well-appointed caravan in which the Schutte brothers Ben and Andrew, proprietors of Empire Amusement Parks, resided. These social occasions were entirely unrelated to the ten-year contract that was subsequently concluded with the Schuttes in 1992 for the right to continue providing the funfair facility at the annual Show![25]

Apart from its new ten-day duration, which included a second Sunday to offer working people an extra opportunity to visit, the 1993 Royal Show had two unusual features. The first was much less visible to the general public than the second. By rotation the Association of Show Societies of Southern Africa held its lengthy annual general meeting and lunch during that Show. Representatives from ten of the country's leading agricultural societies attended to exchange information and Mark Shute was elected the association's next chairman.

More prominently, that year Zulu King Goodwill Zwelithini and one of his wives made a first-ever informal visit and toured the main agricultural exhibits with a small entourage. The occasion was recorded as 'historic and particularly memorable for the courtesy, cordiality and enthusiasm of the Royal couple' which was subsequently re-affirmed in a letter from the King's secretary.[26]

The Queen took a keen interest in the needlework and other parts of the Crafts and Home Industries Section. His Majesty particularly wanted to view the cattle but appeared to be disappointed when introduced to a woman, Iona

Stewart of the Cattle Section committee, to guide him around the stalls. He seemingly put her to the test by requesting her to determine the age of a huge Charolais bull, being unaware of her farming experience and expertise in animal science which was soon to earn her a doctorate.

The King was visibly delighted after she had gently parted the animal's lips to examine its teeth and announced that it was eighteen months old. Then they completed the visit hand-in-hand with his accompanying Queen negotiating the dung pats and mud as best she could in high heels. King Zwelithini insisted that Iona Stewart should attend him again in the event of any further guided tours and later even visited her own farm to inspect her nGuni herd, being a herd owner himself. On that occasion disaster was narrowly averted when she only just managed to prevent her pet goat from butting him!

During his visit the King expressed interest in exhibiting some specimens from his own Royal nGuni herd at the 1994 Show. He was formally invited to do so but did not take up the offer.[27] Instead popular additional features that year included helicopter flips which the Transvaal Helicopter Training Services offered from the nearby Bird Sanctuary area by arrangement with Ben Schutte's Empire Amusement Parks. He, in turn, undertook to pay his share of the proceeds (R5 per flip) to the RAS in lieu of sponsoring the fireworks display.

During the 1980s and early 1990s other sections of the Show were also characterised by a mixture of traditional features and new developments.

The **Horse Section** was able to re-introduce evening jumping sessions from 1987, following the installation of a new holding area for jump equipment and, not least, the previously mentioned upgraded Main Arena floodlighting. It was also fortunate to enjoy the services of talented course designers in Natus Ferreira, Captain Christopher Coldrey and Harold Preston.

In addition, there was a succession of experienced judges in Gavin Doyle, David Stubbs, Rosemary Dorward, Abby Tren, Sue Serritslev, Prudie Dix, Gregory Goss, Vanessa de Quincey, Eric Lette and Jo Jefferson. It could also rely on highly competent arena marshals and numerous efficient jump stewards. Indeed, by the mid-1990s the Horse Section boasted more honorary officials over a longer period than any other section of the Show.

Unbeknown to show jumping competitors the electronic equipment that timed their rounds was not always reliable and, on occasion, a high quality wrist watch had to be used! Bert Cornell eventually constructed a digital timing device using aircraft landing lights to provide a strong enough beam for daytime use. It was his first digital venture and was to last for twenty years before being replaced by a modern digital timing unit using Wi-Fi.

Entries in the showing classes declined in 1985 due partly to the prohibition on cross entries in the riding horse class from the hack and hunter categories, but driving re-appeared at the Show with eleven competitors in three classes. In 1987 the cancellation of the Horse Section due to an equine flu outbreak was only narrowly averted but in most years there was strong competition in all disciplines and classes. This was especially the case in dressage and showing, although entries declined in 1988 due to inclement weather.

The odds and evens and knock-out pairs competitions attracted public interest, as did the interprovincial show jumping team championship and the 1988 South African show jumping championship. That year a tractor-driven vibrator was used to break up the soil of the Main Arena to a depth of approximately 150 mm, followed by the application of fertiliser to ensure adequate future growth and a green finish.

By 1990 it could confidently be asserted that 'the Royal Show continues to provide a larger audience for show jumping than any other venue for the sport in the country'. While the horses were generally well-behaved, the same could not always be said of the free-ranging Jack Russells some owners brought to the Showgrounds and tended to get underfoot in the members' pavilion![28]

The interprovincial show jumping team championship in 1991 and other subsequent big competitions maintained a high level of public interest while substantial improvements to the drainage and surfaces of the new saddling paddock and the Main Arena were undertaken to meet the demands of competitors. William Dreboldt organised the loan of equipment with which to rip the Main Arena surface to a depth of 750 mm preparatory to harrowing, top dressing and fertilising.

This treatment had the desired effect with horse show organisers reporting favourably on the underfoot going as well as the repairs to the water jump. Such events were unfortunately becoming uneconomic primarily due to the cost of acquiring 'free shavings' from distant sources. Organisers of these occasions were henceforth required to buy sufficient straw bales to provide the bedding needed for the Royal Show, the bloodstock sale and all other equestrian events held on the premises.[29]

Following the introduction of breed classes for Arab horses, European Warmbloods and Welsh ponies as well as the introduction of CA-grade show jumping, the overall number of entries in the section came close to the 1970 record of 1 442 with another 1 391 in 1993. That year there were protracted negotiations between the RAS and the Natal Horse Society when the latter declined to construct the sand dressage arena that it required adjacent to the Chatterton Road stables until the former was willing to enter into a formal sub-lease agreement.[30]

During the 1980s and 1990s the Horse Section enjoyed numerous sponsors, some of longstanding. These included Toyota SA, Benson and Hedges (which withdrew in 1985 after twenty years of support), Protea Assurance, Dhoda's Market, LCM Engineering, Shuter & Shooter, Suncrush, the Arab Horse Society (Natal), the Imperial Hotel and the *Natal Witness*.[31]

By the 1990s there were several major awards that enjoyed generous sponsorship. These included the Prix St Georges dressage, the champion show horse, the Natal grand prix, the open show jumping championship of Natal, the young riders' championship of South Africa and the Natal children's grand prix which were sponsored by the likes of Garden City Motors, Tongaat-Hulett, White Horse Distillers, Belgotex, NCD-Clover, Simba and Coca-Cola.

Philip Smith won the Toyota show jumping championship of South Africa in 1984 and the Natal open championship in 1986 for Smith and McCall with Cloud Nine. Barry Taylor won the Natal grand prix in 1986 and the Natal open in 1987 with Fidelity Guards Powerforce for Fidelity Guards. Lynda Kirchmann's Ogden Nash and Mazapan won the champion show horse awards in 1985 and 1989 respectively.

Among other prominent winners in the late 1980s and early 1990s were Vicky Mostert of Panorama Farm at Table Mountain with Aramis and Donner Madeira (dressage), Ginny Fine with Bold Bronze and Leanne Mostert with Aventica (champion show horses). Bruce Dewar won with Connecta Okinawa (centenary grand prix), Anneli Wucherpfenning with Audi Say Hello and Audi Olympian (open show jumping championship of Natal and Total SA accumulator) and Bryce McCall with Blue Rock (Natal grand prix and open show jumping championship of South Africa).

The **Cattle Section** also maintained its reputation in terms of quantity, quality and competitiveness despite the depressed state of the industry in the 1990s. It was grateful to Terry Bailey and his wife for computerising its catalogue, for which he was presented with a PMB 150 medallion. By 1992 the section's committee meeting minutes were being produced on a word processor for the first time.[32]

The new record total entry of 2 115 in the Cattle Section set in 1986 was boosted by 650 specimens from 30 exhibitors in the national championship which the Hereford Breeders' Society of Southern Africa held at the Show. The judge, Gene Wiese from Iowa, described it as one of the most impressive gatherings of the breed that he had seen anywhere in the world. In 1986 and 1987 there were as many as nineteen breeds represented and 128 exhibitors. The Frieslands led the way in the dairy division with 257 entries and twenty exhibitors in response to a 75th anniversary celebration. The following year eighteen breeds were on show with the Charolais national championship as the main feature attracting 185 entries.

The overall increase in entries during the 1980s resulted in congested conditions and an accommodation shortage for attendants and fodder. Jeremy Snaith pointed out that British agricultural shows were investing substantial amounts in upgrading their respective cattle sections. In 1988 a quota system had to be imposed on exhibitors and the commercial cattle were judged on a separate day to make double use of the more than 60 stalls.

Requests by cattle farmers for permission to take their livestock home before the end of each Show were ongoing. This was particularly so in the case of the dairy breeders who pointed out that cattle were only required to remain for three days at the Tulbagh and Zimbabwe Shows.[33] By 1989 the congestion had been further eased by the provision of additional stalls and some rooms for attendants as well as the piping of more water from the Dorpspruit.

The Natal Holstein Friesland Club was eventually permitted to issue its own show catalogue so that its members could select their entries for the Show as late as possible instead of meeting the Society's closing date which was far too early for them. A subsequent improvement, at Iona Stewart's suggestion, was the provision of special pens to allow for the inclusion of unhalter-trained steers in the commercial cattle section to attract new exhibitors and additional entries.[34]

One of the 1990 Show's highlights was the Sussex national championship with Dr Hugh Seton as judge. It attracted 317 entries from 21 exhibitors and was regarded as the largest exhibition of that breed anywhere in the world. In all there were seventeen breeds on show that year amounting to 1 948 entries, just 167 short of the all-time record, including 531 from other beef breeds, 484 in the dual-purpose category, 384 in the dairy breed section, 131 interbreed entries and 101 in the slaughter section. There were 125 exhibitors, sixteen showing Simmentalers and fourteen each showing Brahman and Friesland.

The Standard Bank parade of cattle was followed in the Main Arena by the judging and presentation of awards in the national junior stockman competition which Stephen Vosloo of Duiwelskloof won. By then the Stock Owners parade of commercial cattle had also become a highlight and good quality entrants continued to be attracted thanks, in part, to the ongoing inclusion of the Meat

Board's national carcass competition.

Stock Owners continued throughout each year to assist all the livestock committees and to provide skilled personnel to ensure the smooth operation of their various competitions. Weston Agricultural College's staff and students also continued to contribute to the Show by undertaking numerous duties while the Cedara College's students' cattle judging competition proved successful enough to be expanded.

In response to an ongoing decline in support for the commercial classes Terry Strachan took it upon himself to provide commercial cattle exhibitors with a monetary incentive by augmenting the prices they received when the carcasses were sold. The problem was also remedied, in part, by the introduction in 1992 of a students' challenge class which attracted entries from Cedara, the Universities of Natal and Zululand and from Weston. The involvement of these enthusiastic young participants in this and other classes augured well for the future.

In an effort to foster this interest and at the suggestion of Iona Stewart a Future Farmers Competition was introduced the following year with ten male and female entries between 18 and 25 years of age.[35] Classes for unhalter-trained animals were also introduced in the commercial category and temporary pens were erected to accommodate them. The South African Angus national championship was held again at the Royal Show in 1994 attracting 194 entries from twelve exhibitors while the second Future Farmers and the national junior stockman competitions also attracted interest.

The Cattle, Sheep and Pig sections all enjoyed the sponsorship of Stock Owners Co-operative Limited throughout this period. Other prominent supporters were the Meat Board, Standard Bank, the Tongaat-Hulett Group, Smith Kline Animal Health Products, Renown, Meadow Feeds, Eskort Bacon Co-operative, Pfizer Laboratories and the Natal Agricultural Union. In 1988 the City of Pietermaritzburg presented six carriage clocks to the major sections to celebrate its own centenary.

The Cattle Section boasted eight major awards by 1984. That year J. Muller won the beef supreme championship as well as the grand champion bull and cow categories with Leachman Stall and Leachman Telindalee in the Red Angus national championship. Keir Hall of Mooi River won the grand champion bull category in the Aberdeen Angus national championship and then the RAS Gold Cup in 1986 for the beef supreme champion with Star's Ensign of Epworth and Star's Eclipse of Epworth respectively.

In 1985 a new competition for the best turnout among the cattle stalls had 'an excellent effect'. It was introduced at the suggestion of John Brewitt who donated a trophy for this purpose while Coopers Animal Health subsequently contributed a cash prize. In response to poor support for 'the string of 12 cattle' competition in which each group represented one owner/breeder a new class, 'the string of ten', was introduced. This enabled breed clubs to enter a good representation of their particular breed by selecting animals from all their exhibitors. It eventually resulted in as many as 140 head of cattle, each halter led, being in the ring at one time to produce one of the most spectacular events in the Cattle Section.[36]

C. Mattison won the RAS Gold Cup for the dual-purpose supreme champion in 1985 and the Stock Owners performance test class with Camton Caleb, a Pinzgauer bull. R.M. Greene won the Gold Cup for the beef supreme champion with his Sussex bull Kittle Cattle Nero and the following year won the Sutherland Trophy for the beef reserve supreme champion with Start's Starlight Wizard. John Fowler was particularly successful in 1985, winning the awards for champion Angus bull, supreme champion Angus animal on show and beef reserve supreme champion.

In 1988 W.J. Sharratt exhibited the champion commercial animal and won the Gold Cup for the beef supreme champion with an Angus bull, Red Top

---

J.M. (John) Fowler matriculated at Hilton College before studying at the Royal Agricultural College of Cirencester and then working in the South African fertiliser industry for eighteen years. He launched his farming career by purchasing 500 acres of the family farm at Petrustroom, most of which was inherited by his elder brother. Fowler subsequently bought neighbouring land prior to resigning his job and concentrating full-time from 1977 on developing what was now a viable farming property.

His father had been a longstanding supporter of the RAS and exhibitor of Ayrshire cattle, involving his two sons in the Society as early as 1953 when he acquired life membership for them. John Fowler began exhibiting his own Aberdeen Angus cattle at the Royal Show from the late 1970s and was involved in RAS committees before becoming vice-president (1984–1987) and then president (1987–1991). He was to recall his term in that office with great pleasure, readily giving credit to the contributions made by his vice-presidents, former Rand Show executive member Bruce Lobban and businessman Ron Glaister.[37]

Class of 85, that year and again in 1989 with an Angus cow, Rosewick Lovana Elizabeth. D.J. and V.W. Armitage exhibited two Friesland cows, Weydon Design 1017 and Weydon Simon, to win the RAS Gold Cup for the dairy supreme champion in 1989 and 1990. They won that competition again in 1992 and 1993 with the same Holstein Friesland cow Weydon Cinament 1500.

At the Sussex national championships in 1990, J.L. Orford & Sons exhibited the breed champion bull Redholm Handsome 4th and cow Redholm Exstasy 12th, with the former also winning the beef supreme championship. John Fyvie won the beef reserve supreme championship with a Santa Gertrudis cow Cathrew JF 85-20 in 1992 as well as the dual-purpose supreme championship with a Simmentaler cow Luvale Hafke's Lucienne.

William Angus (Pty) Ltd won the former with a Simmentaler bull Wisp-Will Mandrake 4th in 1991 and 1993 as well as the dual-purpose reserve supreme championship with the same animal in 1994. Mapstone Brothers exhibited the champion commercial animal and champion national carcass in 1992, winning the former title again the following year.

The **Sheep Section** hosted the national junior shepherd competition in 1994 after first being introduced the previous year. During the 1980s the section also needed enlarged facilities as entries increased from less than 300 in the late 1970s to a record 857 in 1990. Some exhibitors contributed pens when national breed championships and the presence of internationally recognised judges like Ray Sharritt from Australia attracted greater interest to the section.

The provision of a new clubhouse added to its prestige and the section was finally deemed to have come of age in 1986 when its supreme championship award was judged at the official opening in the Main Arena together with those of the Cattle Section. It was adjudged 'the best-ever Sheep Show at the Royal with good judging and good competition', including support from the Free State breeders of S.A. Mutton Merinos.[38] There were seven breeds on show amounting to a record 685 total entry while the highlight of the slaughter category was the Meat Board's first annual national lamb carcass competition, which boosted entries to 52 groups involving fourteen exhibitors.

By 1990 temporary sheep pens under canvas had to be erected in spite of an extension of the area provided for permanent accommodation. The Suffolk national championship that year attracted 37 exhibitors with 117 entries while provincial championships were held for Hampshire Down (148 entries), Ile de France (101) and South African Mutton Merino (88). In the interests of improving quality in the annual national lamb carcass competition the slaughter sheep category was delayed to September to coincide with the Society's Select

Ewe Sale.

In 1991 the countrywide recession in the industry was reflected in a 38% decline in entries compared with the record 857 the previous year. The absence of any entries from Ile de France, Landsheep, South African Mutton Merino and Suffolk exhibitors was to some extent compensated for by strong interest in the Border Leicester national championships held at the Show.

A major attraction was the new sheep shearing complex which involved meticulous planning in collaboration with Ian Rutherford and the National Woolgrowers' Association. As the *Annual Report* for that year recorded 'it is a measure of the shearing community's dedication to competition shearing that some 40 judges and officials were on hand throughout the two days at their own expense'. Credit for the construction of the complex was also due to John Ronald of SA Mutton Merino and his sheep shearing sub-committee and to the sponsors Stock Owners, Coopers and Smith Kline with Southern Cross Industries supplying the shearing machines.[39]

After André Pretorius, president of the Natal Wool Growers' Association had opened it, 360 sheep were shorn in front of large audiences before H. Salisbury of Aussie Shear was declared the first Natal and Midlands machine shearing champion, for which he won a trip to Australia. Elliot Ntsombo from Lesotho won the hand shearing competition and again in 1993 and 1994.

In 1993 the National Woolgrowers' Association acknowledged the quality of the complex and of the officials in attendance by awarding its national sheep shearing championships to the Royal Show, the first time it was held outside Bloemfontein.[40] With the title of The Southern Cross-Hoechst Shears on offer, the machine shearing class involved 40 competitors and the hand shearing 29, including one entrant from New Zealand and two from Western Australia. Hoechst, Southern Cross Industries and Voermol Feeds generously supported the competition while Flip de Jager and family of the Ladysmith district provided the necessary 600 Dohne Merino sheep.

As the lean years for mutton and wool producers extended into the 1990s this continued to be reflected in the number of entries attracted to the Show with the Hampshire Down Society providing most of the support. In 1993 that breed's national championship was held at the Royal Show with the twelve exhibitors' 261 entries demonstrating a quality many considered superior to that shown at the nationals of any other breed in the country.

In 1994 there were still only thirteen exhibitors and 231 entries, of which six exhibitors and 101 entries came from the Hampshire Down Society. The shearing competition was, however, again a resounding success with eleven entries in the hand class and fourteen in the machine class while the de Jager family, as before, provided the 450 sheep needed. Southern Cross Industries, Hoechst Animal Health and Voermol Feeds continued their sponsorship.

However, the Sheep Section experienced a decline in its major awards from five in 1984 to two by 1994. Prominent among the repeat winners were Mrs Tim Hancock of Karkloof who exhibited the grand champion ram at the 1984 Hampshire Down national championship, the champion 'pen on the hoof' among slaughter sheep in 1985 and the grand champion ram and ewe in 1988. She also won the Epol Trophy for the supreme champion sheep in 1989 and the same award for the supreme champion and breed champion ram in 1993 and ewe in 1994.

Craigieburn Estates produced the champion 'pen on the hoof' in 1984 as well as the champion lamb carcass that year and again in 1985 before winning the national championship in that category in 1986. D.N. Trollip exhibited the senior champion ewe and grand champion at the 1985 Border Leicester national championship as well as the supreme champion in 1987. In those years Dr and Mrs D.H. Wang produced the senior champion ram and the grand champion ram in that category. Dr D.M. Dommett carried off virtually all the prizes at the 1991 Border Leicester national championships but A. Hardingham's Corriedale Ram was deemed the best sheep on show.

The **Pig Section** had the smallest number of exhibitors of any category during the 1980s and early1990s but compensated for this in enthusiasm and the quality of its entries. Like the Sheep Section it had a reputation for its hospitality to breeders which was facilitated by the completion in 1987 of a new clubhouse where several sponsored functions were held. The parade of champion pigs on bakkies through the Main Arena as part of the opening ceremony was also a popular innovation.[41]

A new judging and sale ring complex as well as additional holding pens were added in time for the 1988 national pig championships. Britain's Lionel Organ of Cheltenham judged the Large Whites on that occasion and a supreme champion award in this category was introduced.

The Meat Board's first national pig carcass competitions which featured in the 1986 and 1987 hook test competitions were not repeated but the total entry of 489 in 1988 necessitated the imposition of quotas for the available accommodation. Although this figure dropped to 319 the following year and 248 in 1990, with Landrace and Large Whites prominent as before, the foundations had been laid for what could become the best pig complex at any South African show.

The national championships at the 1991 Royal Show, with Robert Overend from the Royal Ulster Agricultural Society judging Landrace and Duroc, perked up the number of entries to 279. It also prompted the replacement of the remaining wooden sites with 24 concrete units and saw the introduction of a new interbreed class for the 'champion group of three animals on show'.

This was followed in 1992 with a first appearance of the Chester White breed at the Show although it was only a token entry involving two exhibitors. The following year Adderley Boerdery from the Cape exhibited for the first time which helped to mollify concern about the paucity of exhibitors from outside Natal.

The Pig Section witnessed a drop in its major awards from six to one in the decade after 1984. Roger Barnes, J.H. Meschede, Nico Dorfling and Darrol Hopkins featured as frequent winners in various categories. Baynesfield Estate was prominent in producing champion porkers and in the national carcass competition while Dorfling and Hopkins were still to the fore as competition declined in the 1990s.

Pedigree cattle and pig sales, which by 1988 the Society's auctioneers Stock Owners Co-operative had conducted for 40 years on a non-profit basis,[42] largely reflected the prevailing state of those industries and wider economic conditions. During the 1980s individual prices and overall turnovers were generally good with several records being broken and no less than seven bettered in 1988 alone. The 139 head of cattle sold then attained a record aggregate of R775 100 and a record average price of R5 576. The R20 000 paid for Stephen Muller & Son's Charolais bull also established a Royal Show record.

The pig sale that year similarly achieved a record aggregate of R83 700 for the 139 animals sold. The two boars from Hogan Stud that sold for R5 000 each also constituted a record while the combined turnover for the cattle and pig sales that year of R858 800 more than doubled the record R426 525 achieved in 1987.

More records were broken in 1989 when an average price of R5 690 was attained at the bull sale although the aggregate for the 93 head sold was only R529 150 with little demand for younger animals. Sussex bulls fetched the highest breed average and M.C.N. Orpen of Barkly East sold one for R20 000. The pig sale's turnover was only R56 000 but a new record average of R1 647 was achieved with 34 animals sold and Darrol Hopkins secured the highest price of R5 200 for an untested Landrace boar.

In keeping with declining market trends, the 1990 cattle sale attained a turnover of only R455 400 which was 14% lower than the previous year. Orpen again achieved the highest price of R18 000 for a Sussex bull. In the pig sale Nico Dorfling sold an untested Duroc boar to Ivanhoe Farming for R5 400 although the average price was only R1 148 with a relatively low overall turnover of R37 900.

Market conditions did not improve in 1991 with total turnover in the cattle sales declining to R283 000 and R22 850 in the pig sales in which Duroc and Landrace prices dropped by 30%. Cattle sales declined a further 7.7% the following year and pig sales by 5.3% as the market remained depressed. It was only in 1994 that the former picked up marginally, due largely to the efforts of Stock Owners field officers, with a turnover of R162 750 for 24 head while a Charolais bull attracted the highest price of R7 750.

Pig sales remained in the doldrums as the number of breeders and the perceived benefits of marketing at shows diminished. In 1994 a 'silent sale' was held, involving negotiations between sellers and buyers through auctioneers at the Showgrounds without the formality of a sales ring. Only two animals were sold, so it was unlikely that this experiment would be repeated.

Pig sales had virtually come to an end but before doing so there were some lighter moments. Ron McDonald recalled an informal post-party nocturnal auction at which the successful bidder belatedly realised that he did not have a home for the boar that was now suddenly in his possession. Happily, a suitable haven was found for it in the large piggery at the Hilton College farm.[43]

The **Apiarian (Honey), Dairy Produce, Budgerigar and Cage Bird, Poultry, Pigeon and Rabbit** sections of the Royal Show continued to play a significant role in supplementing its agricultural dimensions by widening the spectrum of its popular appeal. Collectively these sections contributed 3 015 entries to the 1985 Show amounting to almost 50% of the 6 077 livestock total and more than 27% of the 11 016 grand total in all competitive sections. By 1990 they still constituted 38% of the 6 527 livestock entries and nearly 20% of the 12 558 grand total, rising again to almost 50% and 26% respectively in 1994.

The **Natal Beekeepers Association** did not regard May as the ideal time of the year for honey producers to provide exhibits but in 1988 the judge Reg Leveridge, ably assisted by his wife Joan as chief steward, reported on the 'outstanding' quality of the province's honeys. In his report on the 1992 Show he described what had been 'the most enjoyable judging day in my experience due to the outstanding selection of colours and flavours' on exhibition.[44]

The **Dairy Produce Section** had for several years considered itself 'something of a Cinderella', being unable to display some entries including ice cream due to a lack of refrigeration facilities.[45] It was deservedly given greater prominence in 1990 alongside the exhibit of the Dairy Services Organisation with the Natal Fresh Milk Producers Union loyally supporting the Show. In 1991 the number of entries in this section was only 40 short of breaking the 1922 all-time record of 171.

However, by the mid-1990s there was concern about rapidly declining support. This was due to the alarming decrease in the number of manufacturers countrywide following amalgamations and closures, including that of the dairy laboratory in Pietermaritzburg which judged the exhibits. Dairy produce, it seemed, like pig sales had made its last appearance at the Royal Show in 1993. Other sections were faring rather better.

The **Budgerigar Club** drew an enormous 722 entries in 1994 when it hosted the Natal provincial championships. The Natal and Coast Poultry Club similarly attracted breeders from all over the country when it hosted the national championships that year after making its accommodation more appealing in 1985 with renovations at its own expense. The **Natal Rabbit Club** gained greater prominence when its hall was moved alongside Poultry after occupying the same site for nearly 20 years.

Agricultural machinery was another important dimension of the Show and when in 1987 exhibitors in that category expressed dissatisfaction about aspects of it to William Dreboldt, a committee member of the Commerce and Industries Section in which they were included, a meeting was organised to restore their confidence.[46]

The **Crafts and Home Industries Section** was, in Ron McDonald's opinion, indeed 'something of a Cinderella' which had never really been given the full recognition and credit it deserved under a succession of outstanding chairpersons.[47] It still featured its traditional exhibits from members, women's institutes and schools, but from time to time during the 1980s also included a number of honorary exhibits ranging from computers to decoupage. Among these was a display of embroidery from Madeira which the Instituto do Bordado, Tapecarias e Artesanato in Funchal provided after Mark Shute and his wife had visited that island.

Each year the Midlands Woodworkers Guild presented as many as fourteen different craft demonstrations involving a variety of skills which attracted much public interest. By 1988 these demonstrations had expanded to twenty, ranging from lacemaking to the production of lead soldiers. There was a temporary decline in entries from veterans and in the iced cakes and tapestries categories which was attributed to the increasing cost of materials. To some extent this setback was ameliorated by the successful efforts of the ladies committee in raising sponsorships and, importantly, by the increasing numbers of junior and school entries in other categories.

In 1988 the Red Cross Showground crèche was discontinued, which provided the Crafts Section with much-needed additional space following the reduction of its floor area to make room for the NPA Theatre. Other improvements and repositioning of showcases further enhanced the overall presentation of exhibits.

In 1990 this Cinderella section, which Dorothy Robinson chaired, really came of age when it celebrated the 60th anniversary of the construction of what was the Society's biggest exhibition hall, originally built in 1930 for the motor trade. Crafts Hall 60 attracted 5 891 entries which was the second highest total ever and constituted 47% of the 12 558 competitive entries in all categories that year. Scarlet banners decorated the length of the hall which also featured a 1930 model Ford.

The occasion enjoyed strong school support and 35 copies of a special commemorative medal were distributed to longstanding exhibitors and leading winners at a celebratory cocktail party in the members' dining room. The ongoing success of this section, which stood 'head and shoulders' above its counterparts at the Bloemfontein, Cape, Pretoria and Rand shows, owed much to its many voluntary workers and a succession of able leaders including Janet Fraser and, more recently, Dorothy Robinson.[48]

In 1992 junior and school competitors were largely responsible for an overall record 6 272 entries in the Arts and Crafts Section which amounted to 48% of the 13 023 entries in all competitive sections. These were reinforced by the various skilled craft demonstrations and a tableau depicting an early German settler home to coincide with the countrywide German festival that year. Collectively they made the Crafts Hall as popular as ever and drew comparisons with the 1930s when the section was in its heyday.

The record was broken again in 1993 with 6 505 entries, of which nearly 59% were from schools and individual juniors, constituting 51% of the total in all sections. For the first time there were 43 entries from the Sezele organisation involving groups of women living in village communities and it was hoped that their participation could be further encouraged in the future.

The Crafts and Home Industries Section enjoyed the sponsorship of the South African Sugar Association for many years. It was joined, for varying lengths of time, by Consolidated Wool Washing and Processing Mills, Bernina, the South African Knitters Club, Sally Stanley Pleaters and J & P Coats.

Mrs D. Wise won the Championship Floating Trophy (for most points in the senior classes) two years running in 1984 and 1985; as did the Virginia/ Glenashley Women's Institute the Lawson Floating Trophy (for most points in the women's institute competitions) in those years and again in 1987 and 1989. Evelyn Wesley won the former prize in 1986, 1987 and 1989, while the Estcourt Women's Institute won the latter in 1986 and 1988. Sally-Anne Glasspoole won the Mrs Walter Reid Floating Trophy (for most points in the junior classes) in 1988 and 1989. Lorna Oliver won that trophy in 1990, 1991 and 1992, as did Jasna David in 1993 and 1994. Ruth Suttie won the Championship Floating Trophy every year between 1991 and 1994.

The **Commerce and Industries Section** enhanced its public appeal during the 1980s with an increasing variety of exhibits. In 1985 all the available space was filled by 250 exhibitors from all over the country. The aforementioned group exhibit in the Hall of Kitchens was a particularly interesting addition which exhibitors welcomed and it 'heightened the competitive spirit'.[49] Many new applicants had to be turned away as increasing numbers of businesses recognised the trading opportunities the Show offered. The standard of the displays improved noticeably and no less than 36 medallion awards were made, of which new exhibitors won more than 25%.

Recessionary economic conditions reduced the number of new applicants the following year but all stands were again taken although this was not the case at several of the county's other major shows. In the interests of maintaining standards, from time to time some exhibitors were blacklisted and warned about future exclusion if they did not upgrade their displays; six in 1986 and another eleven in 1987.[50]

Food and Nutritional Products (Pty) Ltd made a welcome return after fifteen years while the KwaZulu Finance Corporation, South African Transport Services and the South African Prison Service mounted impressive exhibits. In 1987 new exhibitors Derreg Construction, the Electricity Supply Commission (ESKOM) and the Small Business Development Corporation all won awards.

From the mid-1980s exhibitors readily adopted the stand shell system which Exhibition Stand Systems (Durban) introduced. In 1988 all available sites were again occupied with stand and site lettings exceeding budget while a new Hall 7 and the enlarged Govan Mani building attracted further public interest. So too did the excellent audiovisual presentation with which the South African Transport Services drew 21 500 visitors to the main cafeteria picnic area.

The PMB 150 theme that year was also supported with exhibits by the

Bureau for Information, the City of Durban, the Forestry Branch of the Department for the Environment, the Natal Provincial Museum Services and a medallion-winning entry by the University of Natal's Faculty of Agriculture. The Voortrekker Museum displayed its precious original 1824 Great Trek wagon. This also won a medal and proved so popular that the exhibit was repeated, with variations, at subsequent Shows. In 1990 it attracted more visitors in a week than the museum usually drew in a year.[51]

By the late 1980s the growing retail trade which the Show generated was becoming increasingly evident, as demonstrated by the success of the Hall of Furniture group exhibit. In 1990 there were a record number of exhibitors after additional space was provided following the unfortunate necessity of again declining numerous applicants for accommodation the previous year.

Boere Kooperasie Beperk and the Natal Limousin Club each erected a double-storey pavilion adjacent to the cattle judging ring while the Pietermaritzburg Chamber of Industries presented an industrial group exhibit. This promoted local industry by involving nine manufacturers as well as the industrial development section of the City of Pietermaritzburg. Natal Command presented a comprehensive SADF display while the award of 36 medallions, as in 1985, again attested to the overall quality of the exhibits in the Commerce and Industries Section.

In 1991 prominent exhibitor Malcolmess Ltd participated for the last time but the six exhibitors in Hall 1 attained a new level at the Royal Show when they shared the services of one company to offer a unified design presentation. The Hall of Furniture continued to attract the public as did the recent appearance of the Vintage Tractor Club while the Tongaat-Hulett Group maintained the longstanding quality of its exhibit.

By the mid-1990s all three tiers of government were represented at the Show including the Pietermaritzburg City Council's community awareness project. In 1992 the British Ministry of Agriculture, Fisheries and Food contributed an exhibit which Earl Howe, the parliamentary secretary to the House of Lords, officially opened. It was an indication of South Africa's improving international relations and raised the prospect of further foreign participation in the Show.

In addition, the South African Government occupied the whole of Hall 1 with an exhibit involving sixteen different departments.[52] In 1994 that site featured as the Hall of Technology in which some 250 cellular phone and computer companies promoted their products. Most of them reported good sales figures despite the depressed state of the economy.

Among the Commerce and Industries Section award winners were the City of Pietermaritzburg which secured the Royal Commercial Floating Trophy for the best display in an exhibition hall in 1984 and 1987 while the City of Durban won it in 1988, Pietermaritzburg's centenary year. The Department of Agriculture (Natal region) secured the Royal Industrial Floating Trophy for the best display in an individual building in 1984 and again in 1985. The Natal Parks Board and the KwaZulu Bureau of Natural Resources won it in 1986, 1987, 1989 and 1990 while the *Natal Witness* secured the honours in 1992 and 1994.

The Tongaat-Hulett Group was awarded the Hulett Aluminium Floating Trophy for best display on an open site in 1984 and again in 1986 while Natal Irrigation & Slurry won it for three consecutive years between 1988 and 1990. The SADF secured it in 1991 and 1993, having previously won the Union Castle Floating Trophy, which was the President's award for special endeavour, in 1987. Henderson Truck & Bus won it in 1984 and the Hulett Aluminium Trophy in 1985.

The KwaZulu Finance and Investment Corporation won Die Afrikaanse Sakekamer Pietermaritzburg Wisselskild for the best display by a new exhibitor in 1986 and the Royal Commercial Floating Trophy in 1989, 1990, 1992 and 1993. Kitchen Spectrum secured the former award in 1985 and the Union Castle Trophy in 1989. G. North and Son/Northmec won the new Pietermaritzburg Chamber of Commerce Floating Trophy for the best display of agricultural equipment in 1990 and 1991 while McBean's Implement Co. secured it in 1993 and 1994.

Security became an increasingly important aspect of the Showgrounds development as break-ins and burglaries gathered momentum. Fidelity Guards were contracted to assist and barbed wire was affixed to the perimeter walls. In response to vandalism an electrified fence was subsequently erected around the Chatterton Road stables but when floods swept away the electric barrier under the Station Road bridge it became necessary to revert to traditional foot patrols.

Within the grounds there were some reports of day-time bag snatching and teenage drinking as well as thefts from cloakrooms and vandalism in the public car park but little hooliganism outside the funfair. It was recognised that after-hours access to the grounds was relatively easy via the Dorpspruit. Permanent exhibitors were advised to install burglar bars in their buildings as a large team of security guards, possibly with dogs, would be prohibitively expensive.

Ken Easthorpe found access fairly easy on arrival at the main gate 'about midnight on a very misty night' to deliver a more than 3-metre-tall 'Go Man'

Durban students had made for the Show with polyurethane. When the 'blurry eyed night watchman . . . saw this apparition at the back of the bakkie he took off like a dirty shirt and it was 24 hours before we saw him again'![53]

Physical improvements to the available facilities, like security measures, were ongoing and essential to the continued overall success of each Show. These included the vastly improved floodlighting of the Main Arena. With the exception of the 'hack ring' at the Chatterton Road end, this was now virtually shadow-free, making it possible to reintroduce evening show jumping sessions.

The new floodlights were considered to be 'an excellent improvement', the final cost amounting to R140 751, which was only R751 over budget. It also improved the arena's letting potential for other functions outside Show periods. The junior show jumpers, who were the first to experience the new arena, declared it 'terrific'. The Horse Section also benefited from the acquisition in 1985 of the use of additional land across Chatterton Road, north of the stables, part of which was demarcated for exercising and warm-up purposes.[54]

In addition, the ring road around the Main Arena was reconfigured and improved. More diverse catering services were provided and in 1987 the Stannic Tower replaced the Total Tower. The latter had originally supported the cable for a high wire act in the Main Arena and for twenty years had subsequently served as a landmark and central rendezvous. It was now relocated to the Cattle Section where it was associated with Stock Owners Co-operative and used to make public announcements.[55]

In 1988, on the occasion of its 150th anniversary, the city erected its multi-purpose Victorian-style Council House on what had previously been the Total Tower site. This was to be used for Show exhibitions and hired out for other functions during the year. Through the persistence of Jeremy Snaith's PMB 150 committee, FNB sponsored further improvements to the members' terrace and the Tongaat Group donated the attractive bronze 'family of Cape otters' fountain, which well-known wildlife artist Gordon Cunningham created. The sponsors had rejected his initial 'family of baboons' sketches, but so liked his otters that further castings were commissioned to be displayed at Tongaat properties.[56]

In 1990 it was agreed that the RAS could continue to use the polocrosse fields across Chatterton Road for parking during Royal Show periods at R100 per day. Their ongoing availability for Show time parking nevertheless seemed increasingly uncertain. There was growing concern about the future of members' parking when the NPA transferred the lease of the Standard Cricket Ground across Commercial Road to Voortrekker High School and provided that club with alternative premises.

However, at a City Parks and Recreation committee meeting in 1989 the assurance was given that the property would still be available for this purpose for the Royal, Garden, Book Fair and Echo Business shows. The mayor subsequently confirmed that this would indeed be the case, as did the school board. R6 000 was paid for its use prior to the 1993 Show which formed the basis for a five-year agreement. An additional R2 000 was paid for use at the forthcoming Garden Show.

Cordial relations clearly needed to be maintained with the school as it sought to increase its income. The City Council's attorneys subsequently gave the assurance that the Society's parking rights would be protected by a servitude written into the title deed transferring ownership of the property to Voortrekker High School.[57]

Meanwhile two more permanent improvements were made to the Showgrounds when the new centenary gateway on Commercial Road was opened and the centenary clock tower was unveiled at the cattle judging ring. Total South Africa sponsored the former and FNB the latter, both for the Natal Agricultural Union (NAU) on the occasion of its centenary which it chose to identify with the 1991 Royal Show.

The Union's president Boet Fourie and Kraai van Niekerk, minister of agriculture, toured the grounds on farmers' day which included a members' ox-braai held on open ground opposite the South African Sugar Association's Pavilion. This was made available for a centenary exhibit complete with an appropriate 'seasons of change' façade. The special centenary prizes which the NAU offered in all the livestock classes further underlined the fundamental agricultural dimension of the Royal Show.

During the 1980s the exhibition halls were improved and enlarged, with Hall 4 being expanded by 370m² to provide 21 more stands and increased letting space throughout the year. The addition of Hall 7 made it possible to register a record number of trade exhibits in 1988. The following year a new Honey Hall was erected for the Natal Beekeepers Association and a new house was built on Hyslop Road for grounds manager Don Byres.

New Sheep and Pig Section clubhouses were built under his supervision during the 1980s with generous financial assistance from the Society's auctioneers Stock Owners Co-operative.[58] It also provided a loan for improvements to the Cattle Section, resulting in four new blocks complete with attendant quarters and an additional 61 stalls. Uneven ground in the area was graded for draining and more sections of the Dorpspruit were piped.

The grounds staff used the yellowwood troughs salvaged from the old stalls to make two attractive lecterns. One was for presidential use and visiting speakers at the Showgrounds and the other was presented to the Nottingham Road Farmers' Association on the occasion of its 1987 centenary.[59]

Improvements to the Sheep and Pig sections and the addition of large ablution blocks in the stables area across Chatterton Road were much appreciated by exhibitors and contestants. In 1991 the new 600m^2 sheep shearing complex was opened. Built to the specifications of the National Wool Growers Association, it was expected to attract more pedestrian traffic to what had previously been a backwater adjacent to the Sheep Section.

That year the three bird clubs moved from their congested accommodation in the workshop building into a new R65 000 Bird Hall. This gave their popular exhibits equal prominence with the adjacent Rabbit and Poultry halls and created a logical Fur and Feather Section. It also gifted grounds manager Don Byres more space for equipment and stores (and for the reconstruction of his vintage cars).

In addition to the presidential suite and grill room for members, the public cafeteria which Scott's Caterers' ran next to the Crafts Hall was redeveloped to create an attractive new Show Restaurant to serve the public at events throughout the year. The Imperial Hotel and the Victoria Club retained their catering contracts for the members' dining room/bar and the grill room respectively.[60]

In 1992 the NPA opened its own pavilion next to that of the Natal Parks Board with a special focus on the province's community services that soon became a trophy-winner. Union Flour Mills erected an eye-catching double-storey timber pavilion on the Main Arena ring road to house the Blue Ribbon demonstration kitchen. By 1993 FNB could boast its own pavilion in addition to providing most of the funding for the completion of the members' terrace. Weston Agricultural College developed a particularly attractive farmyard exhibit.

By then all public roads within the grounds had been fully paved. That year Umgeni Water contributed what was described as 'one of the most significant improvements … of the past decade' in the form of the Umgeni Waterway which no other show could match. Its removable dam, boardwalk and floating jetties on the south side of the Dorpspruit between the Stannic Tower and Dhoda's bridge attracted considerable public attention with four commercial enterprises displaying no less than 41 craft.[61]

Strongly motivated by the RAS president, Ron Glaister, phase two of the waterway and Standard Bank's Standard Bankway followed in 1994. This converted a previously unused section of the banks into a 100-metre-long brick-paved pedestrian thoroughfare extending the length of the Crafts Hall. It featured attractive traditional lampposts, nine additional trading cubicles and a pavement café.

Umgeni Water completed its commitment to these developments by establishing its own exhibit overlooking the Dorpspruit. As Glaister put, it 'the overall effect was to provide the public entrance off Commercial Road with a fine new vista that has been shrouded in shrubbery for 90 years'.[62]

Sponsors clearly contributed hugely to the success of the RAS and to the various sections of its annual Show, as well as to the numerous other activities that took place on its premises each year. Prominent among them was FNB which, apart from being the Society's longstanding bankers, sponsored the breakfast at the opening of each Show Week, a tradition which Ian Dixon initiated. As Mark Shute readily acknowledged, without sponsors 'the profitability of the Society would be reduced to the point where it would be virtually impossible for improvements to be undertaken'.[63]

The sponsors, in turn, appreciated the intrinsic importance of these events and also their advertising value. Several of them steadfastly maintained their support through years of economic recession and socio-political uncertainty prior to South Africa's first democratic election in 1994.

During the 1980s South African Breweries, Hulett Aluminium, Tiger Oats and National Milling, Datnis Nissan and Simba Quix were major sponsors of Main Arena displays and official entertainment. By 1990 the Tongaat-Hulett Group, Stock Owners Co-operative, Standard Bank, Tiger Oats and the South African Sugar Association were to the fore. By the mid-1990s FNB was prominent as well as Amabele Breweries, Chubb Holdings, Total South Africa and, as before, Stock Owners, Standard Bank, Simba and Tiger Oats.

As previously mentioned, during the 1980s the Pietermaritzburg City Council, FNB, Hulett Aluminium, the Tongaat-Hulett Group and Stock Owners Co-operative, among others, were all financially involved in effecting permanent physical improvements to the Showgrounds. The RAS was particularly grateful for the many ways in which Stock Owners had assisted it over the years and recognised that 'the Royal was the only major Show that was lucky enough to have such close ties with a co-operative'.

During the early 1990s Stock Owners, Coopers Animal Health and Smith Kline helped to finance the new sheep shearing complex; and FNB the members' terrace. Umgeni Water and Standard Bank were both largely responsible for the development of the Umgeni Waterway and Standard Bankway along the Dorpspruit.

The **finances** of the RAS, its annual Show and the other events that took place on its premises were significantly affected by the availability of generous sponsorships in various forms and by donations. In 1984 the Society's total revenue amounted to R261 058, including a R81 281 surplus on the annual Show, with expenses of R146 963 which reflected an overall surplus of R114 095, the second highest on record.

The 1989 *Annual Report* recorded that it was 'gratifying to see the revenue total pass the R1 million figure at last after 130 Shows in Pietermaritzburg'.[64] More precisely, that year the Royal Show generated R1 160 081 which produced a R336 582 surplus. Thereafter the Society continued to make substantial surpluses on the Show, which rose to as much as R464 227 in 1992, significantly more than the R411 381 drawn from other sources of income that year.

For the first time gate admissions alone exceeded the R1 million mark, amounting to R1 017 866. It was nevertheless recognised that attendance figures tended to fluctuate and efforts to curtail costs were ongoing in the face of inflation, one example being the increasing expense of computerising catalogues and prize lists. In 1993 changes were effected to the RAS rules at the request of the Commissioner of Inland Revenue to enable him to exempt it from taxation in terms of the Income Tax Act.[65]

The Society's financial statement was presented in a different format the following year. Income had climbed to R2 413 985 and expenditure to a substantial R2 403 643 leaving a modest R10 342 surplus. Gate, grandstand and parking receipts contributed R1 151 351 to income, followed by stand rentals amounting to R711 261. Administrative costs were then R1 551 856 (including R1 215 426 in salaries and wages), what were termed 'direct expenses' totalled R574 619 (including R121 017 in prize money) and there was R275 802 in grounds expenses (including special maintenance costs of R53 128).

Current assets valued at R556 259 exceeded liabilities by R304 214 in 1994 while additions to fixed assets during that financial year amounted to a substantial R355 700. It was hoped that, with South Africa's first democratic elections successfully completed, the Society's financial circumstances would significantly improve along with the anticipated economic upswing of the new South Africa.

Moreover, in 1988 the RAS had confirmed a 65-year lease over the Showgrounds. Initially concluded in 1984 for a nominal annual rental of

R500,[66] it included additional land for further stabling on the Chatterton Road sports fields to the east of the existing stable blocks. This long lease obviously gave the Society highly beneficial, although not complete, security of tenure.

The City Council clearly recognised the benefits of the RAS presence in Pietermaritzburg by also continuing to offset its annual rates charges on all the land and buildings with a generous grant-in-aid. However, it would not yet accede to its request to acquire full freehold title to the Showgrounds.[67] That important development still lay in the future.

Membership subscriptions contributed only R159 958 (6.62%) to the Society's total income in 1994, yet the ongoing interest of members was as essential to its survival as ever. There was an increase in membership of nearly 200 in 1984 over the previous year, 4 588 in all. This included a record 3 691 adult members as well as 780 junior members and 117 life members, largely prompted by the 125th Royal Show that year. Thereafter there were fluctuations in all categories, influenced in part by prevailing economic conditions and periodic subscription increases. These were related to rising admission charges prompted by inflation.

In 1986, following legal advice, the executive approved a significant change to the Society's rules to admit 'all races' as members despite the contrary wording of clause 3.1 of Act 36 of 1956. A special general meeting of members unanimously supported this without any discussion in the light of 'changing circumstances' in South Africa and 'to demonstrate goodwill to all sections of the community'. New applications still had to be supported by an existing member of 'not less than five years standing'.[68]

Unfortunately, membership numbers dropped to 4 165 in 1986 but took an upturn in 1989, assisted partly by a new category of competitor members. Their numbers were boosted from 70 that year to 285 by 1994 with the introduction of the three new breed classes in the Horse Section. In 1992 membership rose to 4 629 but declined again in 1994 to 4 512 following subscription and entry fee increases deemed necessary to pay the large rental due on the members' car park.

Other activities that took place, as previously, on the premises during the course of each year provided additional income. They were intended, where possible, to counteract the ravages of inflation but hosting them was also regarded as an important service to the community. The RAS promoted some of them because they were agriculture-related and furthered its declared objectives. Others were non-agricultural but were of financial benefit to worthy causes and the third, most numerous, variety involved hiring out facilities simply as a source of income.

In 1984, two years prior to the change in its membership rules, the Society applied to the Department of Co-operation and Development (Pietermaritzburg) for a permanent permit authorising black persons to attend all shows it organised on its premises. Their presence had been allowed since 1978 but previously the Department had laboriously issued ad hoc permits for each show. The socio-political climate was clearly changing, but the request for a permanent permit still had to be referred to the Department of Community Development in Durban for approval.[69]

By the mid-1980s some requests for the use of the Showgrounds were actually being declined if the anticipated income did not justify the disruption caused to the grounds staff's work schedule between each Royal Show. Even so, there were as many as 325 paying functions during 1994 and by then at least ten buildings were being used on a regular basis each year.

The **Natal Witness Garden Show** was among the more significant of these events. Launched in 1976, it was already a going concern which attracted a number of sponsors for varying lengths of time. These included the South African Sugar Association, South African Breweries, Shuter & Shooter, Tongaat-Hulett Sugar, Carters Garden Centres, Dunrobin Nurseries, Bridgeport Concrete Industries and Suncrush Ltd.

In September 1984 the prevailing weather conditions were much kinder following 'the worst drought in living memory', which had threatened to abort the previous year's event. The 1984 show reflected a pleasing profit of R31 000 but J.B. Poole suggested that it should be held earlier as the 'flowering time for azaleas in Pietermaritzburg was over by the end of September'. They were, after all, the blooms for which the city had become renowned, but he was informed that the dates for the Garden Show were to some extent dictated by 'the flowering time for roses'. The local Rose Society was seemingly quite influential.[70]

Attendance improved by 4 000 to 23 352 but was still 5 000 short of the 1982 record. Guest of honour was Keith Kirsten, the well-known author and TV personality who was then president of the South African Nurserymen's Association. A new attendance record of 28 654 was set in 1985 when the show was extended from three to four days to celebrate its tenth anniversary and to incorporate the national orchid conference and exhibition which filled the Crafts Hall. A record single-day attendance of 11 643 was established, due largely to the Pietermaritzburg Philharmonic Society's evening concert in the Main Arena with a fireworks display accompanying Handel's music.

Thereafter the Garden Show reverted to a three-day format but attendance continued to vary subject partly to prevailing weather conditions. The 1987 event was to be remembered as the flood show, proving to be the wettest on record with 170 mm of rain saturating the venue as the floods of that year began to gather momentum.

In support of Pietermaritzburg's 150th anniversary celebrations fifteen municipalities and other public sector organisations responded to the invitation in 1988 to participate in a co-ordinated display in the Main Hall. The Victoriana exhibit which eighteen garden clubs mounted in the Council House was another strong attraction as was the 10th anniversary floral cake competition.

No less than 150 exhibitors participated in the 1990 Show (compared with the original 64 in 1976) with the central display of the Natal Panel of Floral Art Judges in the Floral Splendour Hall providing a special feature. The disbanding Pietermaritzburg Horticultural Society participated for the last time that year but the Hall of Herbs, Kennel Club dog show and Natal and Midlands Quilters' exhibition were interesting new additions. There were 175 quilts on display in 1991 while another new non-green exhibit was that of the Royal Doulton Collectors Club of Natal.

In 1992 the show was again held over four days with guest of honour Margaret Roberts, the internationally renowned herb expert, delivering the keynote opening address and conducting a well-attended workshop. For the first time medallions were awarded for outstanding commercial and non-commercial exhibits with the Durban Parks Department's Alice in Wonderland display considered the best.

The following year both the Durban and Pietermaritzburg Parks departments presented exhibits of exceptionally high standard and the Pietermaritzburg Philharmonic Society Orchestra performed Handel's Water Music from a platform on the Dorpspruit. Guest of honour Margaret Wasserfall, editor of *Garden and Home*, opened the 'truly magnificent' 1994 show which was based on the theme of children's stories. It attracted many family groups with no admission charge for dependents less than ten years old. A demonstration vegetable garden provided an additional attraction.[71]

The **National Natal Witness Book Fair** launched in November 1984 in the Crafts Hall included the Natal Society Library's storytelling sessions for children. Organised under the chairmanship of Eckhard Oellermann of Shuter & Shooter and with the support of the Associated Booksellers of South Africa (Natal Area), it initially attracted 23 publishers, seven retailers and two libraries.

The event was another first for Pietermaritzburg and drew 4 696 visitors to the well-appointed Crafts Hall that was impressively carpeted throughout and equipped with modular steel shelving units. Book sales were such that the number of stands had to be increased to accommodate 39 exhibitors the following year when there was a 45% improvement in turnover.

This continued to increase while attendance varied although the popular auditorium programme, featuring well-known personalities like Pieter-Dirk Uys, Barry Ronge and Robert Kirby, attracted large audiences. In 1986 it was decided to hold the fair in alternate years, after some difficulty was experienced in filling all the stands due to the state of the economy. It was recognised that, while it was a 'desirable exhibition', the Book Fair would never become 'a major profit earner for the Society'.

There was nevertheless a further 64% increase in turnover two years later. By then publishers reportedly regarded it as 'the best of its type in the country' and by 1990 all 48 stands were occupied although some public resistance to the rising cost of books was detected. In January 1992 it was reluctantly decided not to continue with the Book Fair due partly to disappointing attendance figures but largely to the high cost of hiring the standardised modular steel shelving units which were so essential to exhibitors.[72]

The **Natal Youth Show**, which NCD-Clover, Stock Owners and Taurus sponsored, also originated in the mid-1980s. First organised by Dr Merron Galliers, its objective was essentially educational and was held for the second time in January 1984 and then at that time each year.[73]

The 157 cattle entries in nine classes paraded in the Showgrounds judging ring for prize money which the RAS readily provided. The intention was to develop the knowledge and experience of future farmers as well as demonstrate their ring craft. In 1985 the event featured as part of the International Youth Year and was enlarged to nineteen classes, becoming a two-day event the following year and including a sheep section for the first time.

The Youth Show concept soon spread to other provinces, starting with the Transvaal in 1985 and eventually resulting in the national junior stockman competition. It was justifiably 'a source of pride to the three co-operatives' who had 'pioneered this worthwhile undertaking'.[74] From 1988 'preparation days' were organised for exhibitors prior to the show and a 'junior cattleman' award raised it to interprovincial level with Susan Holderness of Balgowan becoming Natal's first winner.

At the 10th anniversary show in 1992 Karen Gray of Richmond won the senior showman and junior cattleman award for over-13-year-old entrants while the Wartburg Kirchdorf Primary School topped the new inter-school class. The following year Jane Webb of Pietermaritzburg, Claire Evans of Viljoenskroon and Richmond Primary won these three awards while Ian Holderness of Balgowan scored the highest marks for knowledge and preparation. Although the number of exhibitors and entries varied, by the mid-1990s the Natal Youth Show was well-established on the Showgrounds calendar.

The 1980s and 1990s featured several other annual events of varying duration. The **Natal Dairy Sale**, held in the Stock Owners ring, was already well-established after starting in 1976 as the Natal Friesland Club's annual sale. It continued to break its own records for the number of head catalogued, the highest individual sales and aggregate turnovers. In practice it was restricted to Frieslands with the Natal Jersey Club occasionally being involved. By 1987, when the sale celebrated its eleventh anniversary, turnover had increased more than twelve-fold to R201 900 with a record average price of R2 039. Not for the first or last time, that year Melville Oldfield achieved the record individual price for an animal, on this occasion amounting to R6 100.

In 1990 only 50 of the 118 top quality animals offered were sold. Even so, a new record average of R3 622 was achieved for a turnover of R181 100. A.M. Kadodia of Honeydew Farm secured the highest sale price of R9 500 for a Friesland heifer, which was also a record. As a result of stricter screening in 1991 as many as 91% of the high-quality heifers offered were sold with the Taurus trophy being awarded to A.P.D. Turner of Underberg for his lot of heifers which averaged R4 967 a head.

A new record average price of R3 747 and a total turnover of R191 000 was achieved that year with Free State and Transvaal farmers buying 23 of the 51 lots offered. Unfortunately, a decline in the number of head marketed and consequently in sales in 1992 obliged the Dairy Breeders' sales committee to rationalise its provincial programme. As a result, no dairy sale was held at the Showgrounds thereafter. The availability of artificial insemination was making the exhibition of prime livestock increasingly unnecessary.[75]

The **Bloodstock Sale** met a similar fate. Also well-established by 1984, that year the event attracted 639 entries, the highest number of horses ever brought to the Showgrounds. It created a demand for accommodation which was met by building additional boxes and temporarily converting cattle stalls into loose boxes for broodmares.

The Society's sale continued to gather strength, with a turnover of R3.5 million by 1985, but then suffered a severe setback. In 1987 the yearling sale which had been established at Durban's Newmarket stables took place within the same week and deprived it of all offerings in that category. This was despite the fact that the RAS had been selling yearlings since World War II. The result was declining turnovers for two consecutive years although agreement was reached with the Thoroughbred Breeders Association that henceforth all broodmares, weanlings and horses in training would be sold in Pietermaritzburg.

It was poor compensation for the loss of the traditional yearling sales. Chris Smith Bloodstock, auctioneers to the Society since 1979, tried to rebuild the sale without yearlings. In 1989 a pleasing record R5 260 350 turnover was achieved as well as a record average price of R19 850. This seemed to establish the Pietermaritzburg event as the primary broodmare sale in South Africa serving different needs to those in Durban and Germiston.

Instead, it was followed by a steady decline in annual sales attributed to a lack of money in the racing industry. In 1993 only 107 of the 152 horses catalogued were sold for a turnover that was a mere 42% of the 1992 figure. The Thoroughbred Breeders Association took over from Chris Smith Bloodstock as the official auctioneers, but in 1994 resolved to move the mixed bloodstock sale to Durban despite efforts to persuade them to reverse the decision.[76]

The *Echo* **Business Show** was the initiative of C.J. Swart, former president of the Pietermaritzburg Chamber of Industries and the managing director of Pfizer Laboratories. He persuaded the RAS in association with *Echo*, the weekly supplement of the *Natal Witness*, to promote small black business enterprises by holding an event similar to the Matchmaker exhibitions held in Johannesburg and Cape Town.

The first of these filled the Crafts Hall for three days in November 1987. There were 56 stands, 45 of them set up by exhibitors from the informal sector and the rest from the local established business community. Pfizer Laboratories generously underwrote the occasion with some financial assistance from the KwaZulu Finance Corporation, the City Council and local businesses.

The Urban Foundation and the Small Business Development Corporation also lent their support, all of which produced a R11 500 surplus. The Show attracted 5 042 attendees and Petros Magubane, proprietor of the Nkomase supermarket in Mpumalanga, won the *Echo* businessman of the year award while Jomo Sono, football star and entrepreneur was the guest of honour.

The second in this series, held in 1989 against a background of considerable unrest in Natal's township communities, was nevertheless an even greater success and was considered to have served 'as much to bring all sections of the community together in a relaxed atmosphere as it did to create new business for the entrepreneurs.'[77]

No less than 60 small businesses from the informal sector participated, more than 30 established companies subsidised the stand rentals and the presence of Radio Zulu helped to enliven the atmosphere. C.J. Swart continued to chair the organising committee and Don Mkhwanazi, national president of the Black Management Forum, was the guest of honour. Then the event disappeared from the Society's annual calendar, apparently because 'the preponderance of township clothing manufacturers discouraged the organising committee from continuing'.[78]

The **Spring Ewe and Ram Sale** was first held as a separate event in 1988. The Society's Sheep Section committee, in collaboration with Stock Owners Co-operative, arranged it after realising that the Royal Show was not held at a time of the year suitable for the sale of stud sheep. In all, 1 085 head were sold, mostly flock ewes, at an average of R190 and a highest price of R265.

The 1989 sale achieved a somewhat disappointing turnover of R98 482 but it was hoped that the event would establish itself once the currently depressed market improved. Unfortunately, this was not the case and the sale did not survive more than three years.

The **Slaughter Sheep Section** replaced the sale of select ewes in 1991 after it was conceded that turnovers would continue to drop because of the state of the industry. It included the Meat Board's national lamb carcass competition and four sale pen classes for lambs were introduced that attracted the attention of local butchers. Unfortunately, the sale pen auction proved disappointing the following year but the quality of the slaughter sheep was of a pleasingly high standard and that section was re-instated among the Meat Board's top-ten shows in South Africa. In 1993 the winning carcass in the national lamb competition was sold for a record price.

The **Agrimech Exhibition** was arranged for the first time in 1988 by the RAS in partnership with the board of Baynesfield Estate. There had long been a need for an exhibition in Natal focused on the demonstration of hay and silage equipment and crops under field conditions. The event was duly held at Baynesfield Estate on a site immediately to the north of the clubhouse under the auspices of the South African Agricultural Machinery Association, which recommended hay and silage equipment as the main theme that year.

More than 60 manufacturing companies participated with thirteen providing daily machinery demonstrations. Attendance of the farming community was a little disappointing but William Dreboldt, himself an agricultural machinery representative, reported that exhibitors regarded the event as 'more than satisfactory' and the venue 'excellent'. Another Agrimech exhibition was planned for 1991, but did not take place due to the unfavourable economic climate.[79]

The **Timbermech Africa Exhibition**, a bi-annual event, was hosted by the Society for the first time in 1993 after it was attracted from Camps Drift to the Showgrounds due to the latter's superior infrastructure and facilities. The static displays of 100 companies associated with the timber industry were concentrated in the Main Arena as well as adjoining sites and halls while field demonstrations were provided in a municipal plantation in Town Bush Valley.

The exhibition was a great success with Bell the only major supplier to the industry not represented, but with international participation in the form of the Finnish Trade Association and the prospect of even greater national and international interest in the future. Timbermech returned to the Showgrounds in 1995, but then moved to other preferred venues.[80]

The **Angus Bull Sale,** which the Angus Club held in 1994, presented a new challenge in the form of handling unhaltered bulls. This was resolved in the short-term with temporary pens but needed a permanent solution, especially as a Sussex Sale was anticipated for the following year.

Pig and Poultry expos were also successfully held in the Olympia Hall in 1995, 1996, 1998 and 2000 in association with the KwaZulu-Natal Pork Producers' Organisation and the KwaZulu-Natal Poultry Institute. The RAS had been involved in the Rochdale/Hogan Pig Sale as early as 1985.[81]

Administrative and grounds staff, along with an army of largely unsung volunteers, continued as previously to contribute in numerous ways to mounting and running the Society's annual Royal Show as well as the various other activities which took place on its premises during the 1980s and 1990s. Indeed, the RAS relied upon the support not only of its employees, members, exhibitors, officials and judges but also upon that of the media, police, traffic officers and other municipal staff. As general manager Mark Shute put it in 1985: 'The success that the Society is able to achieve and the extent to which it is able to improve its facilities from year to year is very largely due to the unstinted support that it receives from all sections of the community in Natal'.[82]

As before, the RAS owed a huge debt to those who served on its executive and in its committee structure. When in 1987 Ron McDonald's presidential term came to an end long-time executive member Hubert von Klemperer declared that 'the progress of the Society during the six years in question has outstripped that of any previous period in the history of the Society'. He added that 'this was due largely to the sensible, modern business lines' on which it had been run. McDonald and Von Klemperer were both awarded civic honours.

Bert Cornell recalled McDonald's friendly nature and the long hours which he spent on the premises during show time, visiting all the halls, chatting to stall holders, staff members and judges and ensuring that each Show progressed as smoothly as possible.[83] When he was succeeded by the well-known Midlands farmer John Fowler and then by businessman Ron Glaister, a healthy balance in the succession between the representation of agriculture and commerce was maintained.

In July 1992 Glaister hosted an executive think-tank at his home, recognising 'the need for creative planning to give direction to the future activities of the Society'. It was then agreed that a new master plan should be formulated to succeed the one which Professor Paul Connell had undertaken in the 1970s and that it should incorporate the then proposed new exhibition hall and Dorpspruit marina.

Attention was also given to socio-political changes in the country; to the need for more effective marketing, particularly aimed at the Durban area and the black community; to emphasising the agricultural flavour of the Show; and to maintaining contact with livestock and agricultural machinery exhibitors. The raising of loans and use of the grounds for other purposes to improve income was also discussed. So too were the Society's committee and management structure, its grounds staff wages and the need for 'an outward-looking policy' towards the smaller shows in the province and those further afield.

In that regard in 1987 John Fowler had already represented the RAS at the Cape Show before its subsequent closure and the following year had opened the Mooi River Show. In 1996 he opened the Bathurst Show in the Eastern Cape while during the early 1990s the Society was also represented at the Rand, Pretoria and Harare shows.

The last was particularly interesting as an event that was being held within the first decade after independence. As such it provided food for thought with regard to planning for future shows in the new South Africa. The RAS representatives were impressed by the generous hospitality extended to them in Harare and by the popularity of the ethnic craft market there. They were also struck by the racial exclusivity of the cattle judging and adjacent clubhouses as well as the tight security provided at every commercial stand to prevent pilfering.[84]

As far as its own employees were concerned in September 1984 the Society introduced a bus fare allowance for its grounds staff in addition to salary adjustments, subject to periodic review. The general manager's remuneration and employment packages were also revised in 1985–1986 following the closure of his personal company Mark Shute Management (Pty) Ltd which, in terms of his original conditions of employment in 1974, had occupied 20% of his working hours.[85]

In 1985 Pamela Cockle replaced Miss Allen as Mark Shute's secretary. In 1990 Tim Ward joined the RAS staff as assistant general manager and the following year grounds manager Don Byres completed fifteen years of service while Kishore Singh was appointed grounds foreman. The Society lost assistant grounds manager Dave Rutherford and his wife Penny, who had joined the staff in 1988, as well as Pareetha Sirkhot (née Harilall) who was just short of ten years in harness. Miss Rowan replaced Mrs Rutherford and Danie van Wyk retired after sixteen years of service as chief gate steward.

When Tim Ward left, Ian Sumner was appointed late in 1992 from a field of 68 applicants that was reduced to fourteen and then to three through a lengthy process of interviews but he resigned in 1993 for family reasons. Bert Cornell, Marcel Maynard and Ken Easthorpe continued to work efficiently on essential maintenance and other tasks, with the last responsible for balancing the electrical supply over the entire property during Show times. The Society's secretary Jenny van Niekerk resigned in 1988 after her husband's transfer to Durban. but was re-appointed to the staff the following year.

Following Van Niekerk's brief departure Jenny Lovell Greene served as the general manager's secretary from August 1988. That year the conditions of employment, disciplinary and grievance procedures for grounds staff were reviewed along with improved salaries that included the bus allowance. By then it was reckoned that the existing 38-strong grounds staff needed to be increased to 42 to run the Royal Show, excluding artisans, night watchmen and other non-labourers, while the 27 temporary employees should be reduced to 16.

In 1990, after a representative of the Engineering Industrial and Mining Workers' Union of South Africa approached the Society's black staff members, a recognition agreement was concluded using the union's own draft as a basis with the Society's disciplinary, dismissal, grievance and retrenchment clauses attached. It was also in the early 1990s that FNB BOB cards were issued to all grounds staff to minimise the theft of wages in transit. At that time, it was decided to upgrade the existing in-house black staff fund established in 1982 into a non-contributory provident fund for permanent grounds staff members.[87]

In 1993 awards were made in recognition of their long service to tractor driver P.M. Mguni (21 years), charge hand R. Reddy (sixteen years), lawnmower operator G.A. Zuma (sixteen years) and cleaner Sophie Dlamini

J.M. (Jenny) van Niekerk (née Fraser) began a long and important association with the RAS in August 1983 as a shorthand typist and co-ordinator of the Horse Section. From 1989 her portfolio steadily expanded to include virtually every aspect of the Society's administration.

She remembered a pre-computer age when electric typewriters were the latest technology, documents that involved more than twenty copies had to be typed onto stencils, roneo machines were in use, catalogues and prize lists were supplied in draft to local printers for collation, and proof checking was 'a daunting undertaking'. Entries in all sections were recorded manually and members' invoices were handwritten.

In the early 1990s Terry Bailey provided instruction on a single computer with each staff member being allocated an hour a day to practise. Van Niekerk recalled then computerising the membership invoices in her own time, leaving the dot matrix printer to run overnight and collecting the successful end product on a Sunday morning! Computerisation of the stand invoices and all the livestock sections followed. Such was her contribution to the Society's administration over more than three decades that she was accorded honorary life membership long before her eventual emigration to Britain.[86]

(fifteen years). Jean Tilley resigned from the grounds manager's office when her husband was transferred to Johannesburg and Rachel Pretorius replaced her.[88]

Stalwart supporters who were accorded life membership in recognition of services rendered in various capacities included livestock exhibitors R.N. Barnes and T.E.B. (Teb) Hill, fellow Cattle Section supporter John Brewitt, *Natal Witness* chairman D.N. (Desmond) Craib and Frank Mitchell-Innes, all in 1986.

Among those lost to the RAS during the 1980s was Janet Fraser, who died in 1984. She had been a member of the Crafts & Home Industries Section committee since 1950 and was elected its president in 1965 as well as being a member of the Society's executive committee before becoming an honorary life vice-president in 1980. Dora MacGarry, a member of that section's committee since 1969, also died that year. In 1985 Mavis Fowle retired after working as its secretary during the course of fourteen Shows but still served on the committee. Nel Shute continued to maintain the Showgrounds gardens in what Ron McDonald described as 'tip-top shape'.[89]

During the late 1980s RAS honorary life vice-president, Jerry Shepherd died as did Frank and Harry Mitchell-Innes. Jerry had made a huge contribution to the Horse Section, serving as its chairman for seventeen years and was mourned 'by the horse community throughout the Province and beyond'. Frank was an honorary life member who had served on the Society's general committee from 1954 to 1986 and had been associated with the Cattle Section of the Show for many years. Harry had been a member of the general committee from 1955 to 1980 and was a former chairman of Stock Owners Co-operative.

The RAS lost three devoted supporters in 1989. Grace Austen Smith was remembered not only for her 'sterling work for the Produce Section for a period of some 35 years from 1947 until the closing of the Section in 1983' but also for the loyal support she gave her husband when he served as the Society's president between 1965 and 1978.[90] Dooley Filday had contributed hugely to the development of the Commerce and Industries Section while chairing its committee between 1965 and 1978, served on the RAS executive committee for fifteen years and was elected an honorary life vice-president in 1978. Bobby Norton, who was well-known as chairman of the Pietermaritzburg Turf Club, played an important role in developing the Society's annual Bloodstock Sale and had served on the executive committee since 1981.

In 1990 honorary life membership was awarded to Roy Hindle for service to the community after a long career with the NPA and to Monica Reynolds for many years of 'dedicated involvement' in the Crafts and Home Industries Section. That year the RAS lost another executive committee member in Nigel Pinnell. He had served on it since 1976 and from 1965 on the Horse Section committee which he chaired from 1984 in succession to Jerry Shepherd before becoming an honorary life vice-president in 1988.

In 1991 outgoing president John Fowler was nominated as an honorary life president while George Poole, who had chaired the Cattle Section committee since 1978 and served on the executive committee for ten years was nominated an honorary life vice-president. So too was Bruce Lobban for service on the general committee since 1982, as an executive committee member since 1986 and then as vice-president.

Melville Oldfield and Guy Tedder, both of whom had served on the former committee since 1966, were awarded honorary life membership. So too were Fiona Fowler, Bridget James and Naomi Tatham, the latter two for service to the Crafts and Home Industries Section for 40 years.[91] Honorary life member Teb Hill, who died at 80 years of age during the 1992 Show period, had exhibited high-quality Friesland cattle, and more recently Charolais, for more than 50 years at the Royal Show and elsewhere.

The RAS lost two more honorary life vice-presidents in George McIntosh and Jeremy Snaith during the early 1990s, the latter having served on the executive since 1980 and for seven years as a vice-president as well as chairing the 125th anniversary committee. Dr Max Taylor was nominated as honorary life vice-president after serving on the general committee since 1987, on the executive for four years since 1991 and as vice-president for two.

In 1993 the rules of the Society were changed to increase the number of places available for election to the executive committee by making honorary life presidents ex officio members of that body.[92] The Crafts and Home Industries Section lost another longstanding stalwart in Beth Upfold while in 1994 Di Fitzsimons retired after ten years as secretary to that section. She was succeeded by Wendy Parsons but remained on the committee.

The **retirement of Mark Shute** followed at the end of June that year. Following Ian Sumner's resignation, it was decided not to re-advertise the post of assistant general manager/general manager designate and, if possible, to make a suitable appointment from the Pietermaritzburg district through personal contacts. Dr Max Taylor was offered the post, but declined it.

Three candidates, including Paul Inman and Don Byres, were considered before it was offered to W.R. (Rowly) Waller, then director of the Pietermaritzburg Chamber of Industries. Ron Glaister had encouraged him to apply and he subsequently filled the post in September 1993 with ten months to learn the ropes from Shute before his impending retirement.[93]

In 1991 the Pietermaritzburg City Council had already conferred civic honours upon Shute in recognition of the service he had rendered to the Society and its Royal Show. Several farewell RAS functions and an executive dinner were held, at the last of which John Fowler delivered nine verses of appropriate poetry, in honour of Shute and his wife Nel who had worked so hard to maintain the Showgrounds gardens.

Among the gifts they received were honorary life membership, a clock from their Society friends, and a yellowwood bookcase which the president Ron Glaister had made from the troughs in the old cattle stalls. In response, the Shutes presented the RAS with a reprint of the John Tallis map of Natal from 1851, the year the Society was founded.

Since assuming office in 1974 Shute had made a significant contribution towards the completion of numerous additions and upgrades at the Showgrounds under four successive presidents. He had often had to contend with financial constraints, variable weather conditions and regional socio-political tensions. Bert Cornell recalled a very effective working relationship with him that developed into friendship. He also learnt that the best way to suggest possible improvements to the Society's facilities was through Nel Shute who had the knack of persuading her husband that they were his ideas in the first place!

A notable exception was early one morning when, just before the Show was due to open, Cornell and an associate were up in the judges' box overlooking the Main Arena while discussing how the appearance of the premises might be enhanced. They were unaware that a cleaner had inadvertently switched on the microphone, thereby enabling Shute to hear every word in his office which was wired into the system! Nothing disparaging was said and he found the incident highly amusing. Ken Easthorpe recalled a similar occasion when a raucous party held up there at the end of a Show was heard all over the Showgrounds because the public address system was still on! Ron McDonald remembered Shute as a 'great organiser who did a great job'. Echoing the words of Ruth Gordon, Ron Glaister called his departure 'the end of an era', during which time he had exercised 'good judgment and a firm hand' in averting many crises, often behind the scenes.[94]

It was, indeed, a hard act to follow but further challenges and changes lay ahead for the Society, its Royal Show and the new South Africa.

ENDNOTES

1 This chapter makes extensive use of Royal Agricultural Society of Natal (RAS), *Annual Report and Financial Statements* (*AR*) 1984–1995. Gordon, *Natal's Royal Show:* 111.

2 *AR* 1984: 3.

3 Ron McDonald, interview, 12 June 2019.

4 Mark Shute, 'KwaZulu-Natal's Royal Show': 1–2 (unpublished typescript, 2001); Gordon, *Natal's Royal Show*: 125.

5 Bert Cornell, interview, 18 June 2019.

6 RAS, Committee Minutes (RASM), Main Arena Directors, 15 November 1984: 6–7. Ron McDonald, interview, 12 June 2019. Shute 'KwaZulu-Natal's Royal Show': 3.

7 Bert Cornell, interview, 18 June 2019.

8 RASM, Cattle, 27 September 1984: 2, 6. John Fowler, interview, 10 June 2019.

9 RASM, Sheep, 14 September 1984: 2.

10 RASM, Pigs, 6 September 1984: 1.

11 RASM, Apiarian, 23 October 1984: 2; Dairy Produce, 11 October 1984: 1.

12 Shute, 'KwaZulu-Natal's Royal Show': 1–2.

13 RASM, General, 3 September 1985: 2; AGM, 23 October 1985: 5.

14 Shute, 'KwaZulu-Natal's Royal Show': 6.

15 *AR* 1994: 4. RASM, Executive, 14 April 1994: 6 and 30 June 1994: 4; President and Vice-

Presidents, 26 May 1994: 1. RAS, Press Releases, 2 May 1994. *Natal Witness* 12 April 1994.

16 *AR* 1992: 5; 1993: 4. RASM, Arena Directors, 3 October 1991: 5; President and Vice-Presidents, 11 June 1992: 1, 6 July 1992: 3 and 26 May 1994: 1–2. Shute, 'KwaZulu-Natal's Royal Show': 27, 30.

17 RASM, Executive, 29 August 1995: 2–3. Paula Glaister, interview, 7 August 2019.

18 Jenny Fraser, e-mail, 1 July 2019. *AR*s 1984–1995.

19 RASM, Arena Directors, 5 September 1985: 1; President and Vice-Presidents, 16 July 1987: 2. Shute, 'KwaZulu-Natal's Royal Show': 9.

20 RASM, Royal Show Sub-Committee, 11 June 1987: 1. *AR* 1988: 4.

21 RASM, Pmb 150 Show, 3 December 1987: 2.

22 RASM, Commerce and Industries, 19 June 1992: 3; Executive, 28 January 1993: 6; General, 31 March 1994: 6.

23 RAS Press Releases, 20 April 1994. Ken Easthorpe, '62 years' (unpublished typescript): 3.

24 RASM, Executive, 28 January 1986: 6–8.

25 RASM, Arena Directors, 22 November 1988: 4; Executive, 12 March 1992: 6. Ron McDonald, interview, 12 June 2019.

26 *AR* 1993: 4. RASM, President and Vice-Presidents, 3 June 1993: 5.

27 Iona Stewart Personal Papers: 'The beginning of it all' (unpublished typescript). Iona Stewart, interview, 16 January 2018. RASM, General, 29 July 1993: 7; Cattle, 5 August 1993: 4; President and Vice-Presidents, 4 November 1993: 4.

28 Ron McDonald, interview, 12 June 2019. Bert Cornell, interview, 18 June 2019. RASM, Executive, 1 December 1988: 5. *AR* 1990: 6.

29 RASM, Executive, 29 November 1990: 13 and 30 January 1992: 3; Commerce and Industries, 10 September 1991: 9; Horse, 2 October 1991: 8–9; General, 2 April 1992: 2.

30 RASM, President and Vice-Presidents, 4 November 1993: 6; Executive, 10 March 1994: 2.

31 RASM, Horse, 12 October 1984: 1.

32 Shute, 'KwaZulu-Natal's Royal Show': 13, 27.

33 RASM, Cattle, 3 October 1985: 3–4, 22 August 1987: 1–2 and 18 September 1987: 4; President and Vice-Presidents, 31 March 1988: 2.

34 Shute, 'KwaZulu-Natal's Royal Show': 28. RASM, Cattle Executive Sub-Committee, 5 November 1992: 1.

35 Iona Stewart, interviews, 12 February and 29 October 2019.

36 RASM, Cattle, 3 October 1985: 2. Iona Stewart, interview, 29 October 2019.

37 John Fowler, interview, 10 June 2019.

38 RASM, Executive, 3 June 1986: 4.

39 RASM, Sheep Sub-Committee, 6 October 1989: 1–3. *AR* 1991: 6. Shute, 'KwaZulu-Natal's Royal Show': 21–22.

40 RASM, Sheep Shearing, 29 October 1993: 1.

41 RASM, Pigs, 25 August 1987: 1–2.

42 RASM, Meeting with Stock Owners Representatives, 29 July 1988: 1–2.

43 RASM, Pigs, 9 March 1994: 2–5. Ron McDonald, interview, 12 June 2019.

44 RASM, Apiarian, 27 October 1988: 2 and 3 November 1992: 2.

45 RASM, Section Committee Chairmen, 5 April 1989: 3.

46 RASM, Commerce and Industries, 9 September 1987: 3; Executive, 21 October 1987: 3.

47 Ron McDonald, interview, 12 June 2019.

48 Shute, 'KwaZulu-Natal's Royal Show': 7, 17. RASM, Commerce and Industries, 13 July 1989: 5.

49 RASM, Executive, 9 July 1985: 3–4.

50 RASM, Commerce and Industries, 27 August 1986: 3 and 9 September 1987: 2.

51 Bill Guest, *A Fine Band of Farmers Are We! A History of Agricultural Studies in Pietermaritzburg* (Pietermaritzburg: Natal Society Foundation, 2010): 104; Bill Guest, *Trek and Transition: A History of the Msunduzi and Ncome Museums (Incorporating the Voortrekker Complex) 1912–2012* (Pietermaritzburg: Msunduzi and Ncome Museums, 2012): 113–114.

52 Shute, 'KwaZulu-Natal's Royal Show': 25–26. RASM, Sheep Shearing, 4 March 1992: 1.

53 RASM, Executive, 20 November 1984: 6; Gates, 27 March 1985: 5–6; Presidents and Vice-Presidents, 16 May 1986: 1–2. Shute, 'KwaZulu-Natal's Royal Show': 7–8, 10, 21, 26. Easthorpe, '62 Years': 7.

54 RASM, Horse, 19 July 1985: 4, 6; General, 3 September 1985: 3; Executive, 22 April 1986: 1; Main Arena Directors, 26 November 1986: 1–2.

55 RASM, Main Arena Directors, 15 November 1984: 2; Executive, 20 November 1984: 6; Gates, 27 March 1985: 5–6; Presidents and Vice-Presidents, 16 May 1986: 1–2; Cost-Savings, 4 August 1989: 5. Shute, 'KwaZulu-Natal's Royal Show': 7–8, 10, 21, 26.

56 ibid: 13–14.

57 ibid: 16, 29. RASM, Executive, 21 July 1988: 12, 21 July 1989: 5, 24 August 1989: 2, 29 November 1990: 5 and 26 August 1993: 4; General, 1 April 1993: 5–6; President and Vice-Presidents, 16 September 1993: 3–4.

58 RASM, Sheep, 26 June 1985: 6.

59 RASM, Sheep, 14 September 1984: 2. Shute, 'KwaZulu-Natal's Royal Show': 10.

60 ibid: 22, 31. RASM, General, 21 March 1991: 4 and 8 August 1991: 2.

61 *AR* 1993: 5. Shute, 'KwaZulu-Natal's Royal Show': 28. RASM, Section Chairmen, 1 April 1993: 1.

62 RAS, Press Releases, 8 March 1994.

63 *AR* 1990: 5. Iona Stewart Personal Papers: Speech at FNB breakfast, 28 May 2010.

64 *AR* 1989: 8.

65 RASM, Special General Meeting, 7 October 1992: 2–3.

66 RASM, Executive, 5 February 1985: 4 and 2 December 1987: 1; General, 19 March 1985: 3–4.

67 RASM, Executive, 27 January 1994: 4.

68 RASM, Presidents and Vice-Presidents, 18 July 1986: 4–5; Special General Meeting, 15 October 1986: 1–3.

69 Shute, 'KwaZulu-Natal's Royal Show': 45. RASM, President and Vice-Presidents, 15 October 1984: 4, 3 December 1985: 7–8, 15 April 1986: 5; Executive, 30 November 1984: 2.

70 RASM, President and Vice-Presidents, 15 October 1984: 1; AGM, 17 October 1984: 5. *AR* 1984: 7.

71 *AR* 1995: 9.

72 Shute, 'KwaZulu-Natal's Royal Show': 49. RASM, General, 15 October 1986: 3; Executive, 3 December 1986: 7, 1 December 1988: 3 and 30 January 1992: 4.

73 John Fowler, interview, 10 June 2019.

74 *AR* 1987: 8. RASM, General, 3 September 1985: 3. Shute, 'KwaZulu-Natal's Royal Show': 45.

75 John Fowler, interview, 10 June 2019. Shute, 'KwaZulu-Natal's Royal Show': 45–46.

76 RASM, Executive, 20 November 1984: 10, 5 February 1985: 12 and 29 July 1986: 6; General, 19 March 1985: 4–5. Shute, 'KwaZulu-Natal's Royal Show': 48.

77 *AR* 1990: 8. RASM, Executive, 8 September 1987: 3 and 28 January 1988: 1.

78 Shute, 'KwaZulu-Natal's Royal Show': 49. RASM, Presidents and Vice-Presidents, 4 June 1987: 4.

79 RASM, Commerce and Industries, 27 September 1988: 2.

80 RASM, Executive, 29 July 1986: 5 and 1 July 1993: 6. Shute, 'KwaZulu-Natal's Royal Show': 46.

81 RASM, Executive, 19 March 1985: 7. Shute, 'KwaZulu-Natal's Royal Show': 46–47.

82 *AR* 1985: 8.

83 RASM, Executive, 8 September 1987: 4; AGM, 19 October 1988, President's Address: 1. Bert Cornell, interview, 18 June 2019.

84 Shute, 'KwaZulu-Natal's Royal Show': 23, 25. RASM, Executive Think Tank, 25 July 1992: 1–5; AGM, 7 October 1992, President's Address: 3; Commerce and Industries, 22 March 1994: 6. John Fowler, interview, 10 June 2019.

85 RASM, President and Vice-Presidents, 15 October 1984: 4 and 3 December 1985: 7–8; Executive, 30 November 1984: 2.

86 Jenny Fraser, e-mail, 1 July 2019.

87 RASM, Executive, 14 April 1988: 5, 26 January 1989: 6–7, 31 January 1991: 7–8, 7 March 1991: 5 and 28 November 1991: 11–12; General, 25 August 1988: 6; President and Vice-Presidents, 1 February 1989: 1; Show Caterers, 9 March 1989: 5; Rules of the RAS Black Staff Fund, 1 July 1992: 1–4. Shute, 'KwaZulu-Natal's Royal Show': 21, 23, 29.

88 RASM, Executive, 27 January 1994: 10.

89 RASM, AGM, 23 October 1985: 2; Executive, 22 April 1986: 1.

90 *AR* 1989: 9. RASM, Horse, 11 April 1988: 1.

91 RASM, Executive, 10 July 1991: 13; AGM, 16 October 1991: 4–5 and 7 October 1992: 3.

92 RASM, Special General Meeting, 13 October 1993: 2.

93 RASM, Executive, 29 July 1993: 1–2.

94 *AR* 1994: 10. RASM, AGM, 17 October 1990: 5; H.D. Spencer Personal Papers, Derek Spencer Eulogy for Mark Shute, 19 May 2009: 1–3. Ron McDonald, interview, 12 June 2019. Bert Cornell, interview, 18 June 2019. Easthorpe, '62 Years': 2. Shute, 'KwaZulu-Natal's Royal Show': 31–33.

2 'THE LARGEST MIXED EXHIBITION CENTRE IN THE COUNTRY', 1995–2001

IN 1994–1995 THE RAS began to hold a series of think tanks reminiscent of that held in July 1992 at Ron Glaister's home. These were intended to develop new strategies with which to cope with the fresh challenges that the Society faced, including competition from other shows and the emergence of new exhibition centres.[1]

At the same time, it did not lose sight of its commitment to serving the commercial, industrial and service sectors of the community while continuing to maintain the traditional agricultural and home industries dimension of its annual Royal Show. One of the first of these was a livestock think tank held in July 1994 to focus primarily on the needs of the Cattle, Sheep and Pig sections.

Marketing enjoyed a timely new emphasis when Dave Erasmus, promotions manager at the *Natal Witness* and well-known for his role in the Garden Show, assumed the chair of the marketing committee. Clive Hatton coined the slogan 'Bigger, Better, Be There' for the 1995 Show while the committee set out to create greater public awareness through press and radio advertising as well as via the medium of television commercials.

Colourful pictorial banners were erected in the still segregated black, coloured and Indian suburbs of Pietermaritzburg to dispel the notion that the Show was exclusively for whites and to attract more attention. Erasmus argued that a bigger marketing budget was essential to promote the campaign if it was to reach a much broader audience in Durban and elsewhere in the province. Further, that radio was a more effective means of reaching upcountry regions than advertisements in local newspapers.

Di Perrett of Campaign Concepts led a strategy session and produced a draft marketing plan for the RAS. Dave Erasmus conducted a SWOT (strengths, weaknesses, opportunities and threats) analysis of the Royal Show which pointed, among other things, to the need for staff customer-care training, better public relations and more advertising. Close to R156 000 was spent on advertising prior to the 1996 Show and Bob Aldridge, also of the *Natal Witness*, undertook to write the Society's press releases.

The arrival in 1997 of the Show Train at the neighbouring Victoria Railway Siding, which Umgeni Steam Railway organised and the *Daily News* sponsored, proved to be a popular novelty and a useful marketing tool in attracting better future attendance. Thereafter Spoornet undertook to schedule trains every day during the Royal Show.[2]

Attendance at the Show was, as before, of vital importance to the RAS not merely in financial terms but also as a barometer of its standing in the community and the service it was providing. As in previous years attendance was often influenced by the vagaries of the weather. The most unfavourable conditions in twenty years were recorded in 1997 and the best in living memory the following year!

In 1995, for the first time in many years, there was no Ascension Day holiday around which the Show could coalesce. The think tank formed to address this problem resolved to place more emphasis on the two weekends in the Show period as well as on advertising and on arena and other forms of entertainment to attract public interest. It had the desired effect with record attendances being achieved on both weekends that year even though overall attendance dropped. This was possibly due to the distractions of the Rugby World Cup competition and the Comrades Marathon.

The University of Natal undertook a survey during the 1997 Royal Show which advised that future Shows should open at month's end (pay days). It also confirmed that entertainment in its various forms, including the funfair, was the biggest public drawcard. This attracted an estimated 46% of visitors compared to the 4% drawn by the Horse Section and only 2% by the Livestock and Crafts sections respectively. The survey concluded that no less than 23% of visitors came from greater Durban. Despite the understandable reservations of those who feared that the Show might lose its original agricultural focus, not least the farming community, it had to be accepted that pop concerts and other forms of entertainment attracted large numbers and boosted Society finances.

Attendance improved to a record breaking 249 270 in 1998 without the inclusion of any public holidays during the Show period but including the record 59 006 who attended the Main Arena pop festival on the last Saturday evening. Perfect weather conditions, efficient marketing, an excellent entertainment programme and the presence of the SABC all contributed to this success. In the opinion of the president Derek Spencer attendance that year firmly underlined the Royal Show's claim to being KwaZulu-Natal's 'largest single event'.[3]

H.D. (Derek) Spencer's presidential term was from 1995 to 1998. In the mid-1980s he had taken early retirement as shipping officer in the Natal Tanning Extract Company and turned down an offer to assume the post of commercial manager from the South African Wattle Industry. John Fowler subsequently invited him to accept nomination to the RAS executive.

Spencer came to regard it as a disadvantage that he had not 'come through the ranks' by serving in one or other of the Society's sections prior to joining the executive. It was also an advantage that he had no pre-disposed loyalty to any aspect of its activities and was able more easily to appreciate the larger picture.[4]

Public interest in the Royal Show dipped again during the next two years due partly to depressed economic conditions and to the 1999 general election held on day six of that year's Show. It nevertheless remained the biggest single audited annual event in the province. At the turn of the century the RAS still claimed, in the words of its president David Wing, that the Show was 'the largest mixed exhibition centre in the country incorporating a fully-fledged agricultural component'.[5]

Following the example of royal shows in other parts of the Commonwealth, in 2001 a reduced entrance fee was introduced for one day to assist all adult visitors who were physically or otherwise disadvantaged. The 228 813 overall attendance figure that year was the second largest in the Show's history while the highest ever Sunday attendance of 49 644 was achieved with the assistance of the 47 000-odd who attended the Ukhozi FM Festival that day.

Main Arena displays continued to include the now traditional Standard Bank grand parade of animals which opened proceedings and the SANDF's retreat ceremony featuring the Natal Carbineers. Sadly, Nelson Mandela declined an invitation to open the 1995 Show. Equestrian events yet again proved to be a big crowd-puller to the Main Arena with the Defence Force providing popular support in the form of several marching bands.

Suggestions that the hitherto multi-purpose Main Arena might be reserved exclusively for equestrian purposes and preferably not used at all from February to May each year were firmly rejected, not least for financial reasons.

The importance of maintaining it in as good a condition as possible was nevertheless recognised and in 1995 an Aero King machine was purchased to promote deeper, healthy root growth. It proved effective enough despite the presence of an underlying layer of clay.

The Turf Club's Verti-drain was installed but concerns about the grass, drainage and uneven surface were ongoing. In 1996 these deficiencies were compounded by damage inflicted on the arena and jump equipment during a rock concert. This highlighted the problems involved in trying to maintain a multi-purpose facility although efforts to improve the Main Arena continued.[6]

The 1996 Show had to be postponed by a week from the now traditional 24 May–2 June period to 31 May–9 June due to local elections held on 29 May. Appropriately, the chairman of the *Natal Witness* Desmond and Mrs Craib were the guests of honour invited to open the Show in celebration of that publication's 150th year as an independent newspaper.

KwaZulu-Natal's Premier Ben Ngubane was the guest of honour in 1997, with the grand parade being preceded by a drive-past of vintage and modern motor cars as well as tractors. It concluded with a spectacular aerobatic display which the Pitts Special team performed under Shurlok's sponsorship. In 1999 Dr Hermann, executive chairman of Clover South Africa, was the special guest in recognition of that company's centenary and long association with the Royal Show.

The 2000 opening ceremony took place, unusually, in the cattle arena which provided an appropriately more rural ambiance when Lord Vesty, chairman of the Royal Agricultural Society of the Commonwealth (RASC), was the guest of honour. A disconcerting highlight was the escape of a bullock from the cattle parade, which scattered a pink satin-clad women's choir awaiting the late arrival of the Zulu King!

Another much anticipated and now traditional Main Arena event was the annual fireworks performance. For the first time two were held in 1995 in an effort to increase public attendance and both proved successful. The 1999 combined fireworks and laser display set to music was a popular first for Pietermaritzburg and the second the following year was even better received.

The 2001 fireworks were particularly spectacular in celebration of the 150th anniversary of the RAS, although there were complaints that the accompanying music was unduly loud. Earlier objections to noise pollution emanating from the Showgrounds during rock concerts had led to an agreement with Wembley ratepayers and municipal officials concerning the sound level and duration of such events.[7]

Standard Bank and FNB continued to be prominent in sponsoring Main Arena events and official entertainment, along with East Coast Radio, the Simba Group, Stock Owners Co-operative and Radio Lotus. By the turn of the century the Society could boast more than sixty sponsors of its arena and other activities. Indeed, sponsorship became even more important to finance the new think tank strategies developed from 1995 to attract greater public attention.

Additional attractions were generated through the development of a more varied programme of arena displays with the cattle judging ring used as an additional venue. Greater emphasis was now placed on specialised events to draw particular audiences to the Show, such as performances by popular music groups. A 1995 concert featuring Johnny Clegg was not the drawcard that had been expected, but made it possible to assess the popularity of that form of entertainment. Other more successful additional attractions that year were two disc jockeys who appealed to teenagers and the eastern extravaganza Indian members of the marketing committee organised for both Sundays. This included a sari queen competition that was well supported by Radio Lotus.

It was already clear that the new emphasis on various innovative forms of entertainment was popular with the general public when Vodacom sponsored a Soweto String Quartet performance and the 1997 Radio Lotus Bhangra Bash established what was then probably a record arena attendance of approximately 13 000.

Not to be outdone the Ukhozi FM Festival held on the concluding Sunday of the 2001 Show was even more successful with its 47 000 attendees exceeding the 44 000 at the previous year's performance. It was reckoned to be the biggest concert audience featuring local artists ever assembled in the country.

At the first such event a shortage of tickets left gate cashiers at a loss, angry music fans threatening to break down fences and the traffic in surrounding streets gridlocked. Sheets of numbered tickets were hastily printed in the office and distributed to the various gates. A professional company was subsequently contracted to produce all the tickets for exhibitors, Society members and the general public.[8] Another unexpected consequence of the popular last-Sunday concert each year was that in 1998 and 1999 eager attendees crowded the members' grandstand to capacity earlier in the day, necessitating the enforcement of strict exclusion measures.

The introduction in 1998 of a Harrods-style Food Hall was a great additional attraction. Within two years it had become 'possibly the largest dedicated exhibit in the country'.[9] In 2000 an 'agricultural tug of war' competition and herdsmen's braai was added to the annual calendar, the former being well-received by the general public and both by farm employees. A four-day programme of dog displays proved to be both popular and inexpensive. Equestrian events continued to attract public interest during the mid-1990s, but not to the extent of a decade or more previously.

The **Horse Section** still maintained its high standards of yesteryear with, as participant Reeva Wing remembered, the assistance of superb saddling paddock stewards like Rodney Bishop, expert course builders like Harold Preston, Dave Varrie and John Collier, stabling organisers like Lyn Collier as well as judges like Joan Lewis who in 1995 proved to be one of the most impressive ever to appear at the Royal Show.

Bert Cornell recalled that the judges were always 'treated like royalty' with his wife Barbara serving tea in their well-appointed enclosure. She continued to provide this service for some years and subsequently also in the presidential suite. Reeva Wing remembered that in those days the Royal Show was a very prestigious equestrian event which always attracted the cream of South Africa's horse fraternity, fresh from competition at the Rand Show.

According to Jenny van Niekerk the South African National Equestrian Federation rules were very strict with A-Grade show jumpers being required to nominate their classes – medium or national – by 9.00 pm each day. Then she would type up the lists for a local printer to collect an hour later and return the final version in the morning. No changes or mistakes in the nominated order of riding were acceptable. On one occasion, due to an error, the rider who was actually responsible for it threatened her with physical assault and had to be ordered out of the Showgrounds![10]

The Horse Section, as previously indicated, enjoyed sponsorships from at least a dozen quarters. It still attracted many keen contestants for its major prizes. Vicky and Anton Mostert virtually made the Prix St George dressage their own by adding to their earlier triumphs with one or the other winning it every year between 1995 and 2001 on Donner Madeira, Escorial or Ramirez Furi.

Carol Dollery won the champion show horse title in 1995 with Sweet Champion and again in 2000 with Aura of Light as did Di Campbell in 1996 and 1997 with Ruling Party. Miranda Collings was a two-time winner of the Natal open show jumping championship in 1997 and 1998 on Bambix. Winners of the Coca-Cola sponsored children's open show jumping championship of Natal during this period included Samantha Mostert on Arabesque Flashman.

In 1995 the Natal children's grand prix and children's open show jumping championships attracted numerous participants as did the Welsh pony class. The highlight of the show jumping in 1996 was the Swedo Car sponsored Volvo World Cup qualifier which was the first to be held at the Royal Show. Kim Robson won it on Ruby Mountain as well as triumphing in the Clover-sponsored Natal open show jumping championship with the ever-articulate Robin Alexander providing the public address commentary. The Natal children's show jumping championship was again held during the Royal Show as well as the Natal open.

In 1997, in response to waning public enthusiasm and pro-active think tank suggestions, the dressage and show jumping committees successfully injected more public entertainment into their programmes. This included the dressage quadrille, musical kürs and six bar, knock out pairs and ride and drive competitions. In addition, a two-horse carriage brought that year's guest of honour, group chairman of Standard Bank Dr Conrad Strauss, into the Main Arena.

In 1998 Ken Poole won the driving category with Siverstream Carbon and Graphite. Ronnie Lawrence and Avis Vallon Rouge won the Avis World Cup qualifier in front of a capacity crowd after an exciting three-way jump-off. This was held over a course which was the last that Harold Preston designed after 26 years service to the Royal Show. Jennifer Dalton, chair of the showing category, retired after 25 years on the committee.

Another pleasingly large audience attended the Main Arena when in 1999 the Show hosted the South African Pony Club dual discipline international championships. These included the Prince Philip Games, which attracted children and junior entries from all over the Commonwealth. Unfortunately, this necessitated the exclusion of the traditional child and junior classes from the Show that year.

Other concerns had already contributed to the alienation of the Society from the local horse community. The Horse Section 'no longer felt welcome at the Showgrounds' despite the fact that horses had 'always been part of the Royal' and horse owners had always 'felt as if they belonged'. Some riders complained that the surface of the arena had become too hard and was not being maintained to the standards of yesteryear when Max Doepking and Nigel Pinnell took a keen interest in it, ably assisted by two mules which were affectionately known as Max's A-grades.[11] Among other issues the horse committee objected to the executive ruling, following severe damage caused to the Main Arena during a cloudburst, that additional horse shows should cease in the event of wet weather or else the organisers should bear the cost of repairing any consequential damage to the arena.

They were also under the impression that the executive showed no interest in replacing the 96 stables which had been demolished to make way for other developments and that the proposed restructuring of the executive committee would exclude representation of the Horse Section, among others. In response it was officially pointed out that the RAS made no profit from weekend horse events, that all sections of the Show were expected to 'toe the line for the benefit of the whole in the general interests' of the Society's finances and that the Horse Section's expenditure was 'still too high'. Further, that while 'horse [events had] always been and will continue to remain a traditional and valued activity at the Royal' they could no longer be regarded as 'part of the Society's core business'.

Most of the Horse Section's committee followed suit in sympathy when its chairman, the hardworking and dedicated John Plummer, subsequently resigned with some bitterness but for reasons that were not made entirely clear. He was nevertheless to be remembered as a personable individual whose dedication to the interests of that section was unfortunately sometimes at odds with the necessarily broader vision of the RAS management. Thereafter David Wing chaired a so-called horse working group with the intention of finding 'a new way forward' before the committee was subsequently re-instituted.[12] The section did acquire an equestrian clubhouse, complete with offices and bar/ lounge facilities, when the British Petroleum building adjacent to the saddling paddock was completely refurbished.

Junior participation declined in 2000 but adult A- and B-grade entries from all over the country increased and culminated in the Fédération Équestre Internationale (FEI) World Cup Series qualifier which Ronnie Lawrence won on Avis Piroli. It was decided to conclude all horse classes on the second Saturday of the Show to avoid clashing with the Sunday evening concert finale.

In 2001 the section appointed Priscilla Young as its new, highly competent equestrian secretary while Brian Lavery joined Dave Varrie in building the courses for the show jumping classes. Britain's Lucy Killingbeck judged all the age group classes which included 100 adult A-grade horses with riders from Gauteng and KwaZulu-Natal. This culminated in the FEI World Cup qualifier which Heiltje van Tonder won on Greta. The children's show jumping classes were re-introduced that year with most entrants hailing from Gauteng.

The driving classes were as popular as ever but, although the standard was high, dressage entries declined to the point where it was doubtful that the discipline would be included in future. Despite increasing participation in all other categories, the ongoing decline in spectator interest was worrying even though this trend was apparently not unique to South Africa. Bert Cornell, among many others, nostalgically remembered a time when the Main Arena was packed with an enthusiastic audience enjoying the showjumping.[13]

The **Cattle Section** also had to contend with new challenges. Among others, it benefited from the decision taken by the Society's 1994 livestock think tank to find ways of reducing the time in which animals were kept at the Show. While it was not possible to help significantly towards farmers' rising travelling expenses it was suggested that prize money, sponsorships and sales opportunities might be increased while the Stock Owners commission on all sales at the Show, of which the RAS received 3%, might be reduced.

Following disappointing bull sale entries, it was realised that cattle sales needed to be improved by attracting more breed sales like those of the Angus and Sussex varieties, that the existing stalling facilities had to be upgraded, and that halter training would no longer always be possible. In 1995 the Society welcomed the installation of the Holstein Club's Mikrite Parlour facility because it 'not only provides the public with entertainment and education, but also provides all dairy breeders with a modern computer monitored facility for use at all shows, sales and farmers' days held at the Showgrounds'.[14]

The 1996 Show featured the Sussex national championships and the Jersey's 75th anniversary with the latter attracting entries from as far afield as Stellenbosch. The Future Farmers competition, aimed at farm managers, herdsmen and up-and-coming farmers up to the age of 30, placed the emphasis on the judging and presentation of animals as well as showmanship and was coupled with the National Youth Show. That year George Poole retired after nearly twenty years as chair of the section.

By 1997 cattle exhibitors, especially dairy farmers, were experiencing increasing difficulty in leaving their farms for lengthy periods. The new chairperson Iona Stewart proposed a fresh format that would split the cattle showing and judging into categories, such as dairy, commercials, beef, dual purpose, the bull sale and Future Farmers. The intention was to ensure that there were cattle present at the Show for its entire duration but with a shorter period for each category and less pressure on the available stalling.

Among the disadvantages of this format were that there would then be no grand parade in the Main Arena and the camaraderie among exhibitors might be lost. The executive committee favoured the new format, as did Standard Bank which sponsored the supreme champion awards. It was therefore decided to limit the judging of dairy cattle to the first weekend, commercial cattle to

G.W. (George) Poole farmed vegetables and pigs with his father at Fox Hill outside Pietermaritzburg and as a child was regularly taken to the Show. His association with the Cattle Section began in 1951 when, barely out of his teens, he was invited to serve as a steward. His only prior familiarity with cattle was on his uncle's dairy farm in Hayfields. While subsequently employed by the municipality he was granted regular leave during Show time as well as time off to attend meetings during the year.

Poole and fellow Cattle Section stewards Guy Tedder and Melville Oldfield became known as the Three Wise Men. He was also involved in the annual Youth Show, was elected vice-chair of the Cattle Section in 1966 and then chair as well as a member of the executive committee from 1981 and subsequently RAS vice-president. In 1991 he became an honorary life vice-president.

His wife Moyra, who similarly came from a farming background and was brought to the Show as a child, also became involved in the Cattle Section. She continued to assist it and the Youth Show following her husband's death in December 2011. That year he had been actively involved in his 60th Royal Show, having missed only one since 1951 due to surgery! In accordance with his wishes, some of his ashes were buried in the Showgrounds which the family recognised as his second home.[15]

the Monday, dual purpose and beef cattle from Tuesday to Thursday and bull sales to the Friday. This indeed eased the pressure on available facilities while still maintaining a cattle presence for the general public to view.

Previously, exhibitors had been expected to keep their cattle on site for the duration of the Show and for security reasons required a permit to remove their livestock from the premises. Among the serious reasons offered in support of requests for early departure was the increasing incidence of farm attacks, particularly in the absence of owners.

On one occasion it was asserted that snow in the interior was rapidly rendering farm roads impassable! Chair of the Cattle Section Iona Stewart phoned a friend who lived in that region to confirm that the claim was spurious! A highveld couple successfully smuggled two cows prematurely out of the Showgrounds in a horse box as such trailers were frequently passing in and out of the gate. Their absence was subsequently discovered and they were banned from the Show for five years.[16]

Despite the advantages of the new format, in 1998 the Cattle Section reverted to its previous arrangement in response to majority opinion. In common with some other parts of the agricultural sector, it underwent a revival of interest during the late 1990s. There were 1 135 entries in 1998 although this did not compare with the 1 970 a decade earlier, as was the case countrywide. Numbers were boosted by hosting the Dairy Shorthorn national championships, as well as the Jersey, Santa Gertrudis and Simmentaler breed provincial championships. Public interest was further aroused by an exhibition of indigenous and South African composite cattle.

Prior to the 1999 Show various facilities were upgraded and some procedures changed to attract more entrants and ensure a more prominent role for agriculture as in the Show's early days. The RAS paid R400 to out-of-province exhibitors who brought three or more animals and a livestock spectacular was planned for the cattle arena as well as a spit braai for all exhibitors.[17]

There was a welcome increase in local and overberg entries, especially in the Dexter classes. A Vryburg exhibitor, A.B. Butler, entered Red Poll cattle, the first in several years, and won the dual-purpose reserve supreme award for his champion bull. The introduction of classes for the youth resulted in the best-ever attended Future Farmers events.

By 2000 there was a further significant increase in the overall number of entries which reached 800 for the first time in years. However, in 2001 the Cattle Section was confronted with a severe outbreak of foot and mouth disease in the region and all potential exhibitors were circularised to indicate whether or not they intended participating in the Show that year. Following an overwhelmingly positive response that section was held as usual. Although the number of cattle entries was slightly down, farmer attendance continued to increase thereby underlining the importance of maintaining the original agricultural dimension of the Show.

The 2001 Show also hosted the Dairy Shorthorn national championships with Australian James Hill adjudicating and a now traditional spit braai was held for the increasing numbers of farmers and representatives of equipment manufacturers attending the Show. By then the classes which the Society's Future Farmers initiative and the organisers of the Natal Youth Show offered on the first Saturday was also attracting enthusiastic participation.

Competition for the six major awards in the Cattle Section was as keen as ever. Mr and Mrs M.N. Oldfield won the RAS Gold Cup for beef supreme champion with an SA Red Angus cow in 1995 and again in 1998 as well as the Sutherland Trophy for best reserve supreme champion in 1999. Whitehorn

Santas won the latter in 1998 and the former award in 1999, 2000 and 2001 with a Santa Gertrudis cow while H.G. Maree won the Sutherland Trophy in 1995, 1996 and 2001 with a Brahman cow.

D.J. and V.W. Armitage secured the John Simpson Memorial Trophy for the dairy reserve supreme champion in 1996 and 1997 with Holstein Frieslands and Paul Meade won the RAS Gold Cup for dairy supreme champion in 1999 and 2000 with a Holstein cow. Ken Baxter won the RAS Trophy for dual-purpose supreme champion in 1998, 1999 and 2000 with a Simmentaler cow as he had done in 1994 while in 2001 E.L. Barry won that prize with a Simmentaler cow as well as the Meadow Feed Trophy for the dual-purpose reserve supreme champion with a Simmentaler bull.

The **Commercial Cattle Section** similarly enjoyed strong support. The University of Natal won the national carcass competition in 1995 and both the champion carcass and champion commercial animal competitions in 1999. Weston Agricultural College won the champion group award in the former competition in 1995 as it had done the previous year and the latter award in 1997. Voermol then provided R2 000 in sponsorship for the champion carcass competition which J's Harlequin Stud won.

The following year nearly 100 head of cattle were entered in the commercial section with the carcasses on view two days later at the Cato Ridge abattoir where the results were announced and the auction conducted. The trophies and ribbons were presented at a champagne breakfast for buyers and exhibitors.

As a result of changes at the abattoir in 1999 the format of the Royal Show carcass competition had to be altered. The auction was now held after the grand parade and the prizewinning carcasses placed on display in a refrigerated show cabinet next to the Stock Owners Pavilion. The carcass cabinet, together with ribbons, trophies and photographs, was then positioned in the Food Hall to give breeders, buyers and sponsors maximum exposure for the rest of the Show.

In 2000 the Commercial Cattle Section attracted a record number of entries and the following year Voermol and Crafcor's incentives, coupled with the Society subsidy of R2 per kg on the carcass auction prices, helped to attract 173 from the province's leading breeders.[18] In 2001 W. and E. Mapstone won the award for champion commercial animal while the top four carcasses were considered to be among the best exhibited throughout the country in recent years.

Mapstone Bros of Baynesfield also produced the winning carcass, as they had done in 1998 and 2000, realising a record national price of R24.40 per kg. Nearly 30 bidders were attracted to the auction of all exhibition grade carcasses which was held in the Food Hall after adjudication. By then the Royal Show's cattle carcass competitions had become firmly established on South Africa's red meat calendar along with those of the Sheep Section.

The **Sheep and Goat Section**, as its new name implied, included goats for the first time in 1995 as well as the return of Suffolks. The goats proved to be of a high standard and in 1996 S.M. Schiever won the supreme champion reserve with an Enobled Boer Goat ewe. Despite a reduction in KwaZulu-Natal flock numbers, sheep and goat entries actually doubled from 156 to 325 and 40 to 77 respectively between 1996 and 1997.

These figures were boosted by the Hampshire Down national championship held at the Show that year and by greater support from Transvaal and Free State breeders. Dorper breeders celebrated their 10th anniversary at the Show and Mutton Merino sheep made a welcome return after being absent for several years with ten animals on show in 1997.

Entry levels were adversely affected by the depressed condition of the sheep and goat industry although in 1999 the Hampshire Down national championship again came to the Show's assistance and attracted two competitors from Gauteng. Static entry numbers reflected the fact that by 2000 the number of pedigree sheep in KwaZulu-Natal had declined to the 1917 level.

As part of its commitment to bolstering the agricultural sections of the Show the RAS undertook to contribute R5 000 towards securing additional entries provided the Sheep and Goat Section matched that figure over a two-year period.[19] Entries did increase again in 2001, providing some compensation for the decline in the Cattle Section that year.

Competition for the two major awards available was still fierce. A previous winner, Mrs Tim Hancock, secured the 1997 supreme champion reserve title with a Hampshire Down ewe. A new winning name emerged that year when Mrs J. Leimer secured the Epol Floating Trophy for supreme champion with a Hampshire Down ram. She repeated this in 1999 as well as winning the supreme champion reserve award in 1998 and 2000.

Rodney Dorning won the supreme champion award with a Border Leicester ram in 1998 and the supreme champion reserve in 2001 with a Hampshire Down ewe while Chricile lle de France Sheep Stud won the supreme champion award with an lle de France ram and then an lle de France ewe in 2000 and 2001.

The slaughter lamb competition was enhanced by the introduction of a Royal Show roller-mark on all carcasses and good prices were achieved each year at the Cato Ridge abattoir. In 1999 prices were unusually high with the winning carcass ranked fourth in the whole country and realising R91 per kg which was a Show and possibly a national record. By 1999–2000 lamb ranked second among all competitive carcasses nationwide.

A **Smallstock Expo** was introduced in 1995 under John Ronald's able management with all breed clubs being invited to present specimens as well as information about them. Associated commercial exhibitors were also welcome to participate with the Expo being further enhanced by the involvement of emerging and trainee farmers from all over the province who, it was hoped, would benefit from the experience.

As intended, the Smallstock Expo attracted much public attention, particularly with its sheep shearing demonstrations. In 1995 it was decided to limit the championships to a single day in view of the shortage of shearers at that time of the year, outside of Lesotho, and to focus on local shearers. Sam Malgas was declared the machine shearing champion and Anery Mpeka the hand shearing champion. In practice it proved difficult to find and transport suitable competitors out of season and in 1996, with entries declining, the shearing competitions had to be cancelled although there were still demonstrations to attract the public.

The following year South Africa's champion shearer, Zweliwile Elias Hans, performed those and subsequently won the world championship held at Gorey in Ireland. In contrast to the Sheep and Goat Section as a whole, by the turn of the century the Smallstock Expo was attracting exhibitors from as far afield as the Karoo while increasing public interest had elevated the sheep shearing demonstrations to one of the major attractions of the Show.

The Pig Section, like sheep and goats, also suffered from declining entries in the mid-1990s with the result that in 1995 it was decided to hold a small expo in imitation of that launched in the Sheep Section. Breeders provided a suitable variety of pigs to attract public interest and the Natal Pork Producers' Organisation mounted an impressive promotional stand in association with a local butchery.

Efforts to persuade them to revive the Pig Section were unsuccessful, despite the convening of a pig think tank to find ways of doing so, due largely to the fear of cross-infection among animals at large shows. The future, it seemed, lay as it did overseas in pig expositions rather than in poorly supported championship competitions where prize-winning was anyway losing its prestige.[20]

In the absence of any competitive entries that section in effect became part of the ever-popular Smallstock Expo which by the turn of the century had also become known as the **Sheep, Goat and Pig Expo**. Collectively, the Cattle, Sheep and Pig sections did not enjoy the varied sponsorships which the Horse Section attracted but they still had loyal support from FNB, Hoechst Animal Health, Meadow Feed, Stock Owners Co-operative, the Natal Agricultural Union, Standard Bank, Tongaat-Hulett Group and Voermol Feeds.

The other agricultural sections **(Honey, Budgerigar and Cage Birds, Poultry, Pigeon, Rabbit and Agricultural Equipment)** did not enjoy that level of sponsorship but continued to attract more or less the same number of entries during the 1990s, except for Honey. It experienced an overall decline despite the best efforts of the Natal Bee Farmers Association to the contrary. The Honey Hall still maintained public interest with informative features such as its observation hive and live bees, but with sometimes disappointing sales.[21]

In 1997 Budgerigars enjoyed an all-time record 813 entries as Poultry did with 1 008 in 1995. Thereafter the Natal and Coast Poultry Club restricted the number of entries per class per exhibitor to ensure high standards of entry. All these sections continued to be valued as an integral part of the agricultural foundations of the RAS. All of them continued to mount displays of a high standard and of interest and educational value to the public.

It was recognised that ideally exhibitors of agricultural machinery should be grouped together in the agricultural sector of the Showgrounds for the convenience of farmers and the general public. William Dreboldt suggested that, in view of the high stand rentals, companies might be permitted to share sites. In 1997, for the first time in many years, there was a noticeable increase in the representation of firms selling tractors and agricultural equipment, led by Ford, John Deere and Massey Ferguson.

Rentals for space still compared favourably with other shows and sales more than justified the expense. There was some concern among exhibitors that while livestock entries and the presence of bona fide farmers appeared to be dwindling, pop festivals were increasing and the Show was too long.

The Society appointed a committee to address the issue but while it tried to boost the agricultural sections it also emphasised that other events helped to keep it 'in a stable financial position'. Meetings were held with the various agricultural sections and were followed by a marketing campaign directed at farmers and implement manufacturers.

By March 2000 'a concerted effort' had been launched 'to re-establish the Royal as the largest mixed exhibition centre in the country incorporating a fully-fledged agricultural show'. Consequently, the RAS was totally 'committed to seeing a growth and revitalization of the agricultural component of the Royal

Show'. Letters of invitation were sent individually to many farmers, including twenty clients on behalf of each tractor manufacturer.[22]

This was well received by the agricultural sections and the participation of manufacturers continued to grow, with the encouragement of Danie du Toit, chair of the Agricultural Implements Section. By 2001 they were all reporting good business as a consequence of the increasing numbers of farmers attending the Show.

Unfortunately, that year what was described as 'a patent disregard of instructions given the organisers' during the Show led to the local Vintage Tractor Club being requested to conclude its association with the RAS. The Society subsequently undertook to reverse this decision provided the club was willing to abide by its rules and instructions. The latter was still willing to participate in future Shows but decided to relocate to Cedara where the provincial Department of Agriculture offered it attractive new facilities.[23]

The **Crafts and Home Industries Section** received numerous accolades for the high standard of its varied and attractive exhibits. In 1995 Di Fitzsimons retired as secretary of the section but her expertise gained from ten tireless years of service continued to be available in her new capacity as vice-president of Crafts and subsequently as its president.

Like parts of the agricultural sector this section was also experiencing an encouraging resurgence of entries by the 1990s. The various demonstrations held during the course of each Show always attracted public interest but the ethnic craft demonstrations arranged in 1997 unfortunately did not materialise. Apart from exhibits by individual African and Indian entrants there was not much success in attracting those types of displays.

Partly in response to the University of Natal's survey which reflected declining public appeal, the section underwent a major change the following year. A third of the Crafts Hall was occupied by the new 40-stand Food Hall which necessitated a more compact set of presentations and the withdrawal of several entry classes. When the new crowd-pulling Food Hall subsequently occupied that entire venue the smaller and much older Hall 5, which was somewhat derelict and backed onto the railway siding, was beautifully refurbished for occupation by the Crafts and Home Industries Section in time for the 1999 Show.

Entry numbers were still high at 3 282 but compared unfavourably with the 5 625 in 1996. It was feared that only the best exhibits could now be displayed in what was a decidedly smaller venue. However, this was avoided by sacrificing the vegetable stalls, cake-icing and floral arrangements. There was some comfort in the knowledge that the latter two categories had a place in the annual Garden Show. Cooking exhibitions were also significantly reduced, but not bottling displays.

These fairly rapid changes gave rise to understandable resentment among organisers and exhibitors who had invested a great deal of energy and money in developing the Crafts and Home Industries Section and its hall. They were not easily mollified but, under Di Fitzsimons' guidance, its committee made the most of these new circumstances.

In time the natural attrition of women's institute and school entries due to demographic and other social changes was to make the reduced space more manageable. Jeanne Mather recorded that in 2000 spirits were still high with great enthusiasm generated during the fortnight of hectic preparation that preceded the Show, culminating in a lunch at the Station Café for all those involved.[24]

That year there were still 34 schools as well as numerous women's institutes and individuals submitting entries. The section re-affirmed its popularity by conducting a visitor count which indicated that more than 100 000 people passed through its now much smaller exhibition during the course of the Show. The re-introduction in 2001 of several interactive displays and demonstrations attracted further public attention, as did the Station Café in the converted ex-St John's Ambulance building adjacent to the Crafts Hall and close to the railway siding.

Female volunteers ran the Café very successfully although a noticeable decline among those willing to assist the section was a source of concern. On a more positive note, sponsorships increased from only two in 1994 to as many as ten by 1996. These included Coats Tootal, Hirsch's Appliances, Innoxa, Kenwood Home Appliances, Sally Stanley Pleaters, Shuter & Shooter and the *Natal Witness.*

Strong competition for the three major awards offered in this section was ongoing but Ruth Suttie was still able to follow her earlier successes by winning the Championship Floating Trophy (for most points in the senior classes) every year between 1995 and 1998 and again in 2000 and 2001.

Similarly, Miss Jasna David repeated her previous prominence by winning the Mrs Walter Reid Floating Trophy (for most points in the junior classes) in 1995, 1996 and 1997. Another previous winner, Swartkop Valley, retained the Mrs Lawson Floating Trophy (for most points in the women's institute competitions) during that period but in the last year shared it with Klip River and Bridgevale, both of whom won it again in 1999.

The **Commerce and Industries Section** still boasted six major awards for which there was increasingly strong competition. Carphone Natal (Pty) Ltd replicated its 1994 success by winning the Royal Commercial Floating Trophy (for best display in an exhibition hall) in 1995 and in 1996 while Eshayamoya Country, featuring local cottage industries, won it in 1999 and 2000. A previous winner, Natal Witness Printing & Publishing (Pty) Ltd, won the Royal Industrial Floating Trophy (for best display in an individual building) in 1995 while Genfoods (Pty) Ltd was successful in 2000 and again in 2001.

The SANDF was another repeat winner, securing the Hulett Aluminium Floating Trophy (for best display on an open site) in 1995, 1996, 1999 and 2000. The Natal Vintage Tractor and Engine Club triumphed in that category in 1997 and 1998 as well as winning the President's Trophy (for special endeavour) no less than four times in 1995, 1996, 1999 and 2001. The Defence Force won that award in 1997 and 1998, the years in which it did not secure the Hulett Floating Trophy.

Lely Southern Africa won the Pietermaritzburg Chamber of Commerce Floating Trophy (for best display of agricultural equipment) in 1995, 1998 and 2001 which Falcon Equipment also won three times in 1996, 1997 and 1999. Unsurprisingly, a variety of entrants won Die Afrikaanse Sakekamer Pietermaritzburg Wisselskild (for best display by a new exhibitor) ranging from Fresh Produce Sales and Marketing in 1995 to the SABC in 1998, Woodridge Country Estate in 2000 and Capital Kitchens in 2001.

The section welcomed the gradual return of motor car exhibitors in 1995 and the representation of every South African motor car manufacturer for the first time ever at the 1997 Royal Show. Despite the best efforts of the section's chairman Tony Hesp, the Sugar Association could not be persuaded to continue its exhibit and its pavilion was sold to the RAS at an 'advantageous price' of R20 000. The Tongaat Hulett group decided not to have an exhibit in 1996 and was replaced by Hulett Aluminium in a smaller area. It subsequently donated its Showgrounds buildings to the Society.

There was a steady improvement in the quality of entrants in the whole section during the late 1990s. This followed the replacement of a frame lock system of exhibits with the new, more efficient Octanorm stand shell system and the contentious exclusion and transfer of what were described as 'those exhibits more suited to a Sunday Park market'.[25]

In 1995 the section opened the first phase of its Royal Market in the area that had previously accommodated the Pig Section. The subsequent removal of 'flea-market type exhibits' from the main halls to the Royal Market created an additional 50 stands at a time when only 40% of the 170-odd exhibition applicants could be accommodated.

This development was coupled with the introduction of trade days (on the Monday and Tuesday) during each Show when school groups were discouraged from attending. The RAS executive was nevertheless very conscious of the fact that their objective was not only to stage exhibitions and provide entertainment for the general public but also to offer education, 'especially in the field of agriculture' and for the youth in particular.

Some schoolchildren arrived at the Show well controlled by their teachers and 'armed with their writing pads and a multitude of questions', primarily relating to agriculture, to which they were expected to find answers while walking around the Showgrounds. Unfortunately, others were left unsupervised and directionless, sometimes overcrowding exhibition halls to the irritation of other visitors and exhibitors. There were also instances of unattended children as young as seven in the Cattle Section, on occasion 'smacking or prodding' animals and suffering injuries themselves.[26]

Major exhibitors welcomed the new developments because they created less crowded conditions better suited to marketing and receiving significant orders. Moreover, the entries were eventually deemed 'of such a high standard as to create headaches for the judges'![27] By 1998 the quality of the stands in the Commerce and Industries Section was considered to be among the best ever achieved, as reflected by the eight trophies and 29 medallions awarded to its exhibitors that year. The standard of presentations continued to improve and attracted a national exhibit by the government of the United Kingdom which, unfortunately, proved to be disappointing. The further expansion of the Food Hall was intended 'to provide the public with an interesting and educational display covering the entire food chain from farm field to dining table'.[28]

By the turn of the century the iNdlovu Regional Council had undertaken a three-year tenancy of Hall 1 and the section boasted about 350 commercial, industrial and service sector exhibitors. It was one of the few, if not the only one, of its kind in South Africa with a waiting list. In the absence of any additional space, it was resolved that all potential exhibitors who met reasonable criteria should be given the chance by rotation to display their products. In this way, at the risk of displeasing longstanding participants, it would also be possible to avoid public perceptions that the Show was getting stale.

In 2001, although the 'Sunday park market' exhibits had almost entirely disappeared from the main halls, there was still concern about the last two days of each Show which tended, as the RAS president David Wing put it, to 'assume a somewhat different persona to that of a traditional exhibition centre'. He appealed to all stakeholders to bear in mind that the first eight days had very decidedly returned to 'the essence of what constitutes an exhibition' and that they 'should accept the reality, the synergies and the benefits of that final weekend, the success of which contributes to every facet of the Society's activities.' Meanwhile, the flea market, relocated to the new Royal Market, more than justified its continued existence by attracting a growing clientele.[29]

Administrative and Grounds staff, as before, made a vital contribution to all dimensions of RAS business, through their 'loyalty, support and dedication', as did many others who were not in its employ.[30] Despite a reduction in personnel, by 1995 the Society still had six administrative staff, four grounds staff and seven artisans. Important changes were to follow.

That year the administrative staff pension fund and grounds staff provident fund were combined into a single provident fund which was non-contributory but permitted member contributions at 6% of salary which was the same rate at which the RAS contributed for all members. In addition, life insurance cover was taken out for all employees and funeral cover for dependents.[31]

Under Derek Spencer's presidential leadership, it was decided to trim the top-heavy administrative structure. In addition to the management committee (Manco) there was a somewhat bloated general committee. The latter now gave way to a smaller executive committee comprising the president, vice-presidents, honorary life presidents, chairpersons of standing committees, some elected members and, as in the previous general committee, the City Council's representatives.

The 'many talents' of the Society's members, including the general committee, were now to be 'put to better use' in 'specialised committees' such as the one focused on marketing. Spencer himself redrafted the constitution and relevant rules and these amendments were voted into force in October 1995. They were not universally popular but, as he later explained, while change may not be 'altogether pleasing…it is the preservation of the whole that we are earnestly striving for'.[32]

The four-person Manco was now able to make quick decisions when necessary and the general manager's title was changed to director. Rowly Waller's tenure under that name was, however, short-lived. After lengthy discussion in June 1996 about the Society's current condition the executive committee unanimously resolved that the president should indicate to him 'that the needs of the case would best be met if he was to tender his resignation'; further, that the president should discuss with the Manco 'exactly how he should proceed with this task'.

At a subsequent Manco meeting Waller expressed his 'displeasure' at the 'non-disclosure of specific reasons for the Executive Committee being unhappy with his performance'. That body nevertheless confirmed that he would not be reinstated. It was also decided to engage professional advice in filling the vacancy and it was suggested that 'a marketing orientated person' might be appropriate as the new director.[33]

It was an unhappy episode in the history of the RAS although Rowly Waller was to be remembered as a likeable individual who was just not suited to the exacting demands of the job. In 1996, while Sue Richards completed the computerisation of all the accounts, Jenny van Niekerk assumed the headship in her capacity as administrative manager and secretary to the Society.

Among other tasks, she was already responsible for overseeing caterers and commercial trade exhibitors. With the assistance of Tony Hesp, that section's committee chairman, she grasped 'the complexities of letting the stands, haggling with rates, sizes and "gentleman's agreement" discounts' to fill all spaces and confirm the bookings.[34]

In the interim between Rowly Waller's departure and the arrival of his permanent successor, Derek Spencer travelled from his smallholding in the Dargle to spend many hours each day in the RAS offices. He was ably assisted by George Poole, John Plummer and not least by Jenny van Niekerk (who had previously been Derek Spencer's secretary) in keeping the ship afloat. This was at a time when the Society's finances 'were in a parlous state', including a substantial overdraft.

A new post-Mark Shute era in the history of the RAS opened when, in November 1996, after other applicants had been interviewed, Terry Strachan began what was to be a long and significant term of office as director (later CEO).

Once installed in office and assisted by Spencer, David Wing and others, Strachan proceeded to return the fragile financial situation to an even keel. In the process he restored the Society to what he later described as 'a functional, operational and liquid entity'. In retrospect he and Spencer gave each other credit for their success in that regard, though the latter described the former as 'the man on the spot' who gave the RAS 'a good shake-up'. Strachan later conceded that this was probably his most significant achievement in office.

It was a reflection of his success that as early as mid-1997 the Pieter-maritzburg Chamber of Commerce and Industries suggested to the Society that their respective positions of general manager be merged. This proposal

T.D. (Terrence, 'Terry' Duncan) Strachan grew up in Johannesburg and matriculated in 1969 at St. Albans College in Pretoria before undergoing military training, during which he became an infantry instructor at Walvis Bay. He then completed a B.Com (Legal) at the University of the Witwatersrand. Strachan subsequently went into business with his father manufacturing protective equipment, specialising in respirators, welding and riot helmets, the latter being used by the British Army in Northern Ireland.

After twenty years he decided to satisfy his lingering desire to go farming and in 1993 bought a property on the Giant's Castle Road near Mooi River. There he initially produced charcoal but subsequently ran a small Red Angus stud herd and supplied cabbages, cauliflower and broccoli to a firm in Durban. After satisfying his farming appetite for two and a half years he sold the farm, expecting to return to his business network in Johannesburg. Instead, Derek Spencer encouraged him to assume the post of RAS director after inviting him and his wife Carol to visit his home in the Dargle.[35]

was declined. By the end of that year the financial situation was 'more than R100 000 better than budget' and Strachan was formally thanked for 'the tremendous turnaround'.

At the time of his arrival FNB Bank anticipated that the RAS might survive for another 18 months. He found what he regarded as something of 'an old boys' club with vested interest groups' focused only on specific aspects of the Society's activities. To some extent all sections felt that their needs justified special consideration, but they were all now subjected to cost analysis and set budgets. Strachan explained that this was essential to ensure the survival of the RAS.

Careful budgeting and strict controls were not well received in all quarters. The 'use of grounds' concept, initiated during Ron MacDonald's term of office, was more rigorously implemented with Strachan resolved 'to turn the showgrounds into a year-round "living experience"'. This would provide what was to become another important source of income. Sponsorships were also significantly increased, rising to over R300 000 in 1998, the largest amount ever recorded for any Show.[36]

Jenny van Niekerk retained her position as administrative manager but in 1997 grounds manager Don Byres left the Society's employ after twenty years of loyal service. K. Singh assumed the role of what was called grounds superintendent and was officially thanked for the 'dedicated manner' in which he had assumed his new post. During the 1998 Show Terry Strachan was taken seriously ill and had to undergo heart surgery, but had recovered sufficiently to attend meetings from late July.

That year, when Derek Spencer's presidential term came to an end, John Plummer, who Spencer had initially nominated, was expected to succeed him. However, that apparently did not meet with universal approval. For this and/ or other personal reasons Plummer suddenly resigned as vice-president and from all committee work including that of his beloved Horse Section. Instead, David Wing succeeded Spencer and inherited the competent administrative team which his predecessor referred to as the Royal Family.[37]

After the Society had completed a further stressful but necessary restructuring of its staff complement there were a few changes, with Carmen Paul and Wendy Burnard replacing Sue Kinghorn (Commerce and Industries) and Sue Richards (accountant) in 2000–2001.

Security continued, as ever, to be an ongoing challenge. Despite escalating crime rates countrywide RAS security personnel, in collaboration with the South African Police and its equestrian unit in particular, kept theft in and around the Showgrounds to a minimum.

In 1998, after it was estimated that as many as 30 000 people entered the Show illegally each year, a more efficient access control system, similar to that of the Rand Show, was entrusted to HW Management. Members, judges, stewards and caterers were offered laminated identity cards which allowed for unlimited entry. An access control committee was formed to monitor the success of the new system which effectively curtailed gatecrashers and reduced the swapping of badges and exhibitors' tickets.[38]

That year and again in 1999 not a single motor vehicle was stolen from an RAS-controlled area but unfortunately one was lost in 2000. By 2001 the perimeter boundaries had been fully walled, save only for the Chatterton Road stables. Following protracted negotiations with Spoornet, a lease agreement was signed to provide parking in the area between the Crafts Hall and the railway siding beyond it. Security was of major concern not only at the annual Royal Show but also at all the other events held on the premises during the course of each year.

The **Natal Witness Garden and Leisure Show** was pre-eminent among these other activities. It now reverted to a three-day format while the standard of its exhibits continued to improve thanks in part to the increasing role played by Dave Erasmus of the *Natal Witness* in its planning and to a vigorous committee chaired by the likes of Else Schreiner and Lolly Stuart.

The show was attended by prominent visitors including British landscape designer David Stevens in 1995 and garden designer Lady Mary Keen the following year. Generous sponsors included Berry's Lawnmowers and Pools, Huletts Sugar, Starke Ayres and Umgeni Water. The show was followed by a think tank and SWOT analysis to consider ways of improving it and attracting more sponsorship.

It was felt that a plateau had been reached with regard to attendance and that it was essential to retain the exhibitor base provided by the Durban and Pietermaritzburg Parks departments as well as to continue attracting the public with different high-quality exhibits. A leisure component was proposed, comprising 4x4 vehicles and camping equipment to attract a new type of exhibitor and a new audience.[39]

In 1996 the theme of the 21st Witness Garden Show, 'In the mood for celebration', was appropriate as it coincided with the 150th anniversary of the *Natal Witness*. Pat Glass of Camelot Nurseries designed the keynote Efekto-sponsored show garden while Cowan and van der Riet Landscapes put together the *Garden and Home* show garden in the Parks Hall. Local landscaper Jo-Anne Hilliar teamed up with Timberlink and artist Shirley Ronald to create the innovative Kiddies Jungle Book Theme Garden and the Bonsai Society produced a large Japanese garden.

Several new refinements were subsequently included with additional prizes and sponsorships as the show became more commercial. In 1997 the floral cake icing competition celebrated its twentieth anniversary and a further leisure dimension was added with the theme 'Gardening for pleasure, time out for leisure'. Every effort was made to produce the best ever show with such novelties as helicopter flips over the city and a Forsdicks-supplied Landrover providing river rides in the Dorpspruit. Unfortunately, hot weather seriously reduced attendance but Chelsea Flower Show designer Ray Hudson declared that the quality of the displays was 'equal to exhibits anywhere in the world'.

In 1998 attendance improved by 6% to reach 21 509 while better access control increased revenue by 23%. By then the Garden and Leisure Show was established not only as South Africa's oldest but also its largest green or horticultural event, attracting exhibitors and visitors from far afield and focusing on the consumer rather than the trade. By then it was also probably the country's largest three-day exhibition of any kind.

In 1999 attendance increased further to 26 372 and the following year 2 000 attended the newly introduced preview evening in aid of Hospice. However, the highest total attendance (at a four-day garden show) was still the 28 654 attracted in 1985 and the highest in a single day the 11 643 achieved that year. By 2001 the show was still attracting well over 20 000 visitors, including many from Gauteng.

That year, after apparently 'irreconcilable' differences of opinion, the RAS parted company from its traditional sponsor of 25 years, the *Natal Witness*. It was 'probably the longest running' sponsorship that the Society had ever attracted. As early as 1997, Garden and Leisure Show committee members had suggested that a 'national' rather than a 'regional marketing campaign' had become 'imperative' if the show was to develop. It was also felt that the *Natal Witness* should play the role of a naming-right sponsor rather than an organiser or partner as the show was actually the property of the RAS.[40]

Instead, the Society entered a three-year naming-right sponsorship agreement with Independent Newspapers, the *Sunday Tribune* being the naming-rights sponsor. Henceforth the annual event was to be known as the Sunday Tribune Garden and Leisure Show. Apart from financial considerations this had the attraction of assuring nation-wide media exposure via several newspapers.[41]

Other activities held at the Showgrounds ensured that its facilities were in use almost every day of each year thanks, in part, to vigorous marketing. These included horse shows, pig and poultry exhibitions, conferences, religious gatherings and social functions including weddings and parties.

Timbermech Africa again held its bi-annual show at the Showgrounds in 1995 which was the biggest exhibition that industry had held in South Africa and attracted interest from all over the country as well as from abroad. The **Bloodstock Sale** of mares and weanlings held in July that year seemed to herald the return of that annual event to Pietermaritzburg but it was not to be. **Equestrian events** included the Mercedes-Benz junior national championships held in 1995 and 1997, as well as the national dressage championships in 1998 and several other important competitions which the Natal Horse Society subsequently hosted. **Angus Cattle and Beef Master Sales** were successfully held in 1995 and 1998, as were the **Natal Jersey Sale**, a **Pig and Poultry Expo**, **Biztech 2000** and the **Bike Fair** in 2000–2001.

Natal and National Youth shows were held in 1998 and 2000 in addition to their participation in the Royal Shows. The Youth shows were sponsored, as before, by Stock Owners Co-operative and Taurus Co-operative, as well as by FNB, Meadow Feeds, National Co-operative Dairies, the RAS and Standard Bank.

The active involvement and service to the community that the Society provided brought with it a heavy administrative load but also financial complications necessitating the containment of expenditure and, where possible, the maximisation of revenue. The other activities held on its premises were often also of significant financial benefit and were to become even more so in the future.

Finance was, of necessity, always a primary pre-occupation of the RAS executive committee. Before the end of 1994 discussions were being held with representatives of the Natal Agricultural Union and the Stock Owners Co-operative with a view to establishing their head offices and those of other agriculture-related organisations in a rented office block at the Showgrounds.

The possible re-siting of the funfair to facilitate this move was also discussed with obvious financial as well as other possible benefits for the Society. Ben Schutte of Empire Amusement Parks welcomed the prospect of relocating his business beyond the stables in Chatterton Road where a larger area was available. It was suggested that the office block project might even include a restaurant and city lodge-type accommodation which would eventually generate sufficient capital to build a conference facility and provide year-round income for the RAS.

It was a step towards launching what was to be called the Royal Mews Trust to manage such commercial developments. The proposal encountered a setback when, in January 1996, Stock Owners advised that they would not be relocating their headquarters to the Showgrounds. Later that year the possible relocation of the Victoria Club was also discussed.

Fortunately, the decline in gate takings at the 1995 Royal Show was more than compensated for by income from the Garden Show and Timbermech, resulting in a R150 000 revenue increase over the previous year. Expenditure on maintenance and security increased by R40 000 and much more was spent on advertising and entertainment as part of the new strategy to improve the Show's public appeal.

This was met by staff reductions, less expenditure on overtime and by contracting out cleaning and refuse removal. Payment of the final R80 000 installment of the Standard Bankway improved the cash flow and facilitated a R192 000 increase in capital expenditure on Showgrounds improvements.

Those expenses had to be cut to a minimum after the 1996 accounts reflected a disappointing R66 628 loss. The Society was becoming unhealthily dependent upon its R400 000 FNB overdraft facility with interest payable at prime rate. At that time no less than five RAS committees were exercising spending power: management, executive, finance, capital projects and marketing.[42]

In the face of a mounting financial crisis, a careful reassessment of every aspect of the Society's activities had to be undertaken under the guidance of Derek Spencer and Terry Strachan. Spencer had already declared that he was not willing to 'lumber the Royal with a large financial millstone'.

This had the desired effect with expenses increasing by less than 1% and total revenue by nearly 20% in 1997. It was also due in no small measure to a 275% increase in sponsorship income after greater efforts were made to improve corporate contributions. As a result, income exceeded expenditure by R513 315, excluding approximately R300 000 set aside for Showground flood control. At the AGM in October 1997 Terry Strachan and 'all concerned' were warmly thanked for their contribution to the Society's financial turnaround.

In 1998, for the first time, the RAS did not once run into overdraft and all its financial records were favourably broken. This resulted in a pleasing surplus of R975 424 prior to transfers being made to reserves for maintenance projects. Sponsorships amounted to R385 000 due largely to the contributions of Clover South Africa, Coca-Cola, Hulett Aluminium, the SABC and Vodacom. It constituted 8% of income with gate takings contributing 46% and site rentals in the Showgrounds 28%.

Salaries and wages were still the largest expense amounting to a manageable 34% compared to 42% prior to the staff restructuring. However, in 1998 municipal rates were levied for the first time in the Society's history. Previously the municipality's annual charges had been met by its own grant to the RAS, but the latter now seemed to be regarded if not as an outdated colonial relic then certainly as a potential additional source of income. In Terry Strachan's view there was no evidence of any appreciation of the contribution that the Society made to Pietermaritzburg's image and in attracting visitors and additional expenditure to it.[43]

The situation necessitated further discipline to ensure that financial resources were not overburdened. Consequently, the 1999 financial results were almost on a par with those of the previous year. Record sponsorships amounting to R421 000 (9% of income) were attributable primarily to corporate involvement, with the Natal Building Society joining the established contributors. As before, the RAS was grateful to all its sponsors, large and small, for enabling it to offer prize incentives to competitors and a widening variety of entertainment to the general public at its annual Shows.

Terry Strachan stressed that 'sponsorship must be viewed as a partnership with the maximum exposure being afforded to the company to ensure their willingness to return the following year'. Corporate sponsorship was accorded to those companies which contributed R70 000 or more. In return their names appeared in all forms of advertising and they were invited to participate in structuring the entertainment and media programmes for the Shows they supported.[44] By 1999 advertising amounted to R300 000 and arena displays R350 000. Despite expensive capital projects for the first time in many years the Society found itself without any long-term liabilities.

Unfortunately, following a longstanding cordial relationship with Voortrekker High School there was disagreement as to what constituted a fair rental for the use of its playing fields opposite Gate 1 off Commercial Road as a car park. In 1997 the RAS made a one-off payment of R4 000 towards the upgrading of those fields. In view of the servitude which it held over the property it could point out to the school that 'whilst the arrangement is not satisfactory for either party, both sides will in perpetuity and in the spirit of good neighbourliness, be compelled to accommodate one another's requirements'.

The Society did acquire the use of the Town Hill Hospital grounds but only for bus parking and for logistical purposes. It also subsequently leased the area between the Crafts Hall and the Victoria Siding from Spoornet and clung to its use of the parking grounds across Chatterton Road. It did so by stressing to the civic authorities its importance to the Royal Show and the economic benefits which it as well as the Garden Show and other Showgrounds events brought to Pietermaritzburg.

When in 1998 Voortrekker and the RAS again could not reach agreement on rentals and profit sharing with regard to parking, the latter eventually took the matter to the High Court for arbitration. It won its case and a rental amount was settled in April 1999 that was close to its original offer. It was to pay R11 500 for the Show that year with an annual consumer price index (CPI) increase thereafter, plus 20% of net income from the Garden Show while the school was to pay all the Society's legal costs.

In 2001 a one-way traffic flow was implemented (in at Commercial Road and out at the other end of the school grounds) while the advent of a new headmaster and school board was followed by what Terry Strachan described as 'a most harmonious and friendly relationship'. The school would eventually assume the administration and control of that parking facility for their own substantial financial benefit, but without any cost to the RAS which was glad to be relieved of it but retained its right of servitude.[45]

The Society suffered another setback in 2000 when it was confronted with an unexpectedly high rates account of R64 000 which, with electricity and water costs, escalated from 1% to 5% of its expenditure. Nevertheless, its year-end balance sheet proved to be the best ever, reflecting a surplus of R1 116 425 prior to transfers to maintenance and bursary reserves.

This was further improved upon in 2001, when the RAS recorded 'the finest set of results in our history'[46] with an impressive net profit of R1 215 734 prior to fund transfers to specific reserves. By then gate takings and parking amounted to 43% of income compared with 2% from membership subscriptions which had declined from 5% in 1999.

Membership of the Society had indeed continued to decline even though attendance at the annual Show was again increasing. Although membership subscriptions clearly did not have a major impact on the financial situation and similar organisations worldwide were suffering the same trend it was still a source of concern. This was attributed in part to deaths, old age and, in South Africa's case, emigration and changing demographic patterns.[47]

In 1999, in an attempt to add value to membership, innovations were introduced such as free teas on the terrace and a successful members' dinner dance. The 1994 membership figure of 4 512 had nevertheless declined to 2 927 by 2001. This no longer included competitor members but, even so, adult members had declined from 3 584 to 2 444, juniors from 547 to 398 and life members from 285 to 85.

Stalwart supporters were still not in short supply from the membership to serve voluntarily on the various RAS committees and in numerous other capacities, although active participation was flagging in some areas.

In 1994 Reed Howes was elected an honorary life vice-president after many years of service, in particular to the Sheep Section. The following year honorary life membership was accorded to Willem Burger for his contribution to the Pig Section, as well as to Neil MacGillivray and to Edward and Barbara Merrick. As previously mentioned, Paula Glaister was also honoured while her husband Ron was accorded an honorary life presidency.[48]

The Society suffered the loss of several ordinary members and of both previous and current office bearers, including its senior honorary life president D.H. White Cooper who died in 1996. The following year Dr and Mrs Chris Saunders were awarded honorary life membership for their support over several years, as were Robin Alexander, Colonel and Mrs Peter Francis, Peggy Hill, Dawn Mackenzie, Joan Pinnell, Dorothy Robinson and Audrey Shepherd.

So too was Jean Griffin in 1998 when outgoing president Derek Spencer and his wife were elected honorary life president and honorary life member respectively. Harry Tully and Harold Preston were also granted honorary life membership as well as Val Field, the retiring CEO of Stock Owners. The following year the Society lost a highly active and long serving member in Hubert von Klemperer.

Hubert von Klemperer was born in 1914 in Germany and educated in Dresden and Munich before emigrating in 1936 to South Africa and serving in the Union Defence Force during World War II. In 1948 when he joined the family business, Standard Yoke and Timber Mills, his membership of the RAS began. He served on its Commerce and Industries committee but a passion for horse riding soon led to his involvement as arena course assistant, course builder and show jumping judge.

He subsequently became the longest serving member of the Society's executive and Horse Section committee as well as an honorary life vice-president. Highly popular among staff members, Bert Cornell remembered him as 'a wonderful old man' who regarded the judges' box 'as his mansion' being 'a world-class show-jumping judge'.

He was indeed recognised at national and international level, chaired the Natal Horse Society and was president of the National Equestrian Federation. Cornell also recalled how every morning during Show time Von Klemperer would announce his arrival on the gate intercom, which was connected to the judges' box.[49]

In 1999 RAS vice-president and long-time general and executive committee member W.C. (William) Dreboldt died suddenly after a lengthy career with G. North & Son and G.C. Tillage, both suppliers of agricultural equipment. His dedication to the Society and infectious enthusiasm for its annual Show was typical of so many who were associated with it in one way or another. As he often said 'at Show time all roads lead to Pietermaritzburg'!

The following year the RAS lost Ron Fowle, another of its longest serving members and a Bird Section office bearer. He was awarded honorary life membership (posthumously). Honorary life membership was also accorded to Mavis Fowle, Mrs P. Wilby, A.J. Howard and Major-General le Roux for services rendered, as well as to Mrs St Clare Peterson, Elizabeth Roberts and Sylvia Moyle. In 2001 Maureen Hurt, June Anderson and Pat Baverstock, who had all served in the Crafts and Home Industries Section, joined their ranks.[50]

Physical improvements to the Showgrounds were ongoing during the 1990s. By 1995 there was concern that the Umgeni Waterway was being under-utilised because it was not in the normal traffic flow. An attempt to secure paddle boats from the Wild Coast Sun was unsuccessful and various other forms of entertainment, including a casino, canoeing demonstrations and radio-controlled boats were considered.[51]

The Showground gardens were all reconfigured and upgraded under the expert guidance of Gordon Stewart and Mike McDonald. Hall 5, the oldest of all, was magnificently renovated. In 1999 the façade of the new Crafts and Home Industries Hall was completed thereby enhancing the premises' reception area. The first equestrian clubhouse was opened and a new horse lane was developed between the Sheep Section and the Dorpspruit.

The Society's conference room was refurbished to provide a more comfortable venue for meetings, a new computer system was installed and the RAS acquired an e-mail address. It was, however, unsuccessful in objecting to the municipality's proposals for the commercial development of the Chatterton Road and Bird Sanctuary sites to the east and north-east of the Showgrounds.

In 1996 the Society offered its own detailed proposals for the future of the development node stretching from the Corobrik site to Voortrekker High School. These suggestions included the regional shopping centre planned for Corobrik as a means of increasing the city's rates base, enhancing its credentials as a conference centre, attracting shoppers and eliminating a serious source of pollution. However, they proposed that the rest of the node should be reserved entirely as a recreational green belt rather than a commercial zone. The RAS even offered to extend its own stewardship to manage the whole area for 'the benefit of the community'.

The Society also continued, without success, to pursue the possibility of converting its lease on the Showgrounds to outright ownership. Derek Spencer expressed his discomfort about 'the funding thereof' at a time when the RAS had no reserves and was still living 'from hand to mouth' with its overdraft facilities increased from R400 000 to R500 000.

A committee was nevertheless formed to consider the feasibility of this proposal which, it was estimated, could cost as much as R600 000. Meanwhile, the Society did submit a successful application to the Transitional Local Council to rezone that part of the Showgrounds adjacent to Chatterton Road, inclusive of the Victoria railway siding at its south-eastern extremity and up to the traffic circle to the north-east.

It also gave some attention to the proposal to develop a value centre in the Royal Market and workshop area of the Showgrounds. Similar centres were already emerging in Durban and promised to generate income by providing warehouse-type premises for rent. However, the currently available 4 000m² was deemed probably too small to accommodate the mix of tenants required to attract a sufficient volume of shoppers.[52]

During 1996 the perennial problem of flooding came to the fore. This was attributed in part to the relatively low level of the Showgrounds (the cattle ring in particular) but also to inadequate drainage associated with property developments up the hill, including Redlands, Woodgrove and Greys Hospital, in what had previously been regarded as a green belt. It was further aggravated by blockages caused by debris washed downstream during heavy rains.

Following consultation with the City Engineer's Department, the situation was alleviated to some extent by removing two old stone bridges, creating a sump in the river bed and clearing blocked water pipes. In December that year there was further severe flooding of the cattle arena when 63 mm of rain fell in less than 30 minutes. The existing water pipes were unable to cope with such heavy downpours from the hill above and the breed clubs were no longer able to secure insurance for their clubhouses around the cattle arena. It was estimated that it would cost R2 million to overcome the problem.

As this was beyond the Society's means less expensive options were considered. The cattle arena and surrounding clubhouses were flooded again in February 1998 when 48 mm of rain was recorded in half an hour. At last, in 1999, in consultation with the City Engineer's Department, the problem of flooding was, seemingly, entirely overcome with the removal of piping and the construction of a carefully planned R360 000 control canal. A storm in 2000 established that it could cope with 60 mm of rain in 50 minutes.[53]

That year animal loading ramps and related facilities were upgraded prior to the Show and a new workshop was completed adjacent to Gate 5. It was also decided to demolish the building which accommodated a pub known as the French Tart adjacent to the new Crafts Hall after it was damaged by fire.

In 2001 McDonald's Garden Centre and an O' Hagans Group franchise opened for business on the land that had been rezoned in the funfair vicinity on the eastern side of the Showgrounds adjacent to Gate 7. Ron McDonald's daughter-in-law Jane designed the Garden Centre which was built on the floor of what had once been the Pig Section. The Society itself demolished 96 stables and invested R1 350 000 in constructing the restaurant. Its financial situation was such that this and the flood control canal could both be paid for in cash.

The **Royal Mews Trust** had been registered by the RAS in September 1998 to serve as a conduit for this taxable commercial development with the president and three honorary life presidents as trustees. Following amendments to its constitution the intention was also to protect its own tax-free status, which was confirmed by the Department of Inland Revenue late in 1997.

In order to maintain this status, it was not to stand surety for the Trust, nor levy it an administrative fee and was expected to charge market-related interest on any one-off loan that it might make to the Trust. In this way it was hoped that the Society might retain its 'not for profit' agricultural dimension and reduce the risks inherent in focusing its financial resources exclusively upon hosting exhibitions while still making an investment that would assist in meeting its rising costs.[54]

For a time, the restaurant proved to be popular while the 30-year lease agreement with McDonalds Garden Centre effectively resulted in the dilapidated south-eastern corner of the Showgrounds being refurbished at no cost to the Society.

The **150th Anniversary** of the RAS was celebrated in 2001, at a time of considerable adversity. Amid both international and regional disturbances there was some understandable concern for the future. The airborne destruction of New York's World Trade Centre, political instability in Zimbabwe, the demise for various reasons of several other agricultural societies and the continuing devaluation of the rand all had a negative impact on business confidence.

To compound the situation, the Natal Midlands suffered a severe outbreak of foot and mouth disease after an absence of several years. The Society made its facilities freely available to the local farming community in an effort to combat the scourge but was unable to prevent the decline in livestock entries at the Show that year. In addition, within a mere two weeks severe weather conditions caused the loss of more than 6 000 head of cattle in KwaZulu-Natal alone, including top-quality breeding stock.

Not least farm attacks, which had been ongoing since 1990, continued to disrupt local rural communities. Nearly 200 guests attended the opening of a small garden in the Showgrounds, near Sappi House, which the RAS dedicated to the memory of those farmers, their employees and family members who had lost their lives. The Dean of the Anglican Cathedral, Fred Pitout, delivered the blessing.

The Society itself suffered setbacks during the course of 2001. The relocation of the Police Equestrian Unit to new premises necessitated the hiring of commercial security for the Chatterton Road stables where, tragically, there was a fire that resulted in the death of a four-year-old boy. Another child suffered serious injury when disembarking from the funfair roller coaster at that year's Show, during the course of which the RAS van was hijacked.

In addition to declining membership there were also unusually large expenses with which to contend. These included an R18 000 loss incurred in co-hosting an unsuccessful gospel concert with a private events organisation. More seriously, the Society was presented with a first-ever rates account, amounting to R194 880, on the municipal property it leased. A permanent 50% rebate in place of a grant-in-aid was still under negotiation. It also had to find R250 000 with which to replace its entire reticulation system to minimise persistent Showgrounds water losses.

Despite all these adverse circumstances there was reason at the time of its 150th anniversary for the RAS to celebrate. A tie was designed for members as a memento of the event and Keith Kirsten displayed four new dream roses at that year's Garden and Leisure Show in honour of the anniversary.

Although membership numbers were declining the Society's website attracted increasing public attention. Its 1 350 pages were completed in 2000 making it one of the largest sites in South Africa. Between May and June that year it had already attracted 16 110 hits. Moreover, in 2001, as previously indicated, the Society had enjoyed the best financial results in its history despite having to meet unexpected expenses.

While the Witwatersrand Agricultural Society went into liquidation, the RAS was in the happy position of being able to re-affirm its commitment to the local agricultural sector. It did so by offering R16 000 in the form of two bursaries to local students involved in agricultural studies. This attracted 31 applicants from whom Christopher Dlamini and Jon de Guisti were selected.

The president, David Wing, was able to report that, with 'a prudent and professional management team in place', the RAS had experienced 'a most successful run' during the previous five years. This had been due in no small part to his own contribution as an accountant in sustaining the efforts of Derek Spencer and Terry Strachan to steer its financial ship into calmer waters.[55]

In March 2001 the Society had the honour of hosting the 19th Commonwealth Agricultural Conference at very short notice. It was held at the Elangeni Hotel in Durban with more than 200 delegates. They included the president of the RASC, HRH the Duke of Edinburgh, who the British Government had advised against visiting Zimbabwe where the conference was initially intended to be held. Reeva Wing recalled that conversing with Prince Philip was one of the highlights of her husband David's presidential term of office.[57]

Also in attendance was the Secretary-General of the Commonwealth, the Rt Hon. Don McKinnon, as well as agriculturists and representatives of international government agencies from throughout the English-speaking

D.A. (David) Wing was president of the RAS from 1998 to 2003. He was a former Horse Section committee member and show jumping judge who had joined the executive in 1997. His wife Reeva had been a showing and dressage participant in horse shows for some years prior to their marriage. As a non-rider he became involved in an administrative capacity, not only in that section of the Society but also at provincial and national level.

Reeva Wing became a national judge while their sons and a daughter-in-law maintained the family tradition of involvement as active participants. Following his term in office which Reeva Wing remembers as 'a very pleasant experience' working with 'dedicated people', David Wing was a valued member of the finance committee (2004–2006) as well as an honorary life president until his death in 2013.[56]

world. On that occasion immediate past president of the RAS Derek Spencer had the honour of being appointed deputy president of the RASC. In that capacity he kept close company with His Royal Highness for the duration of the proceedings and was responsible for officially bidding him farewell at the airport.

Generously sponsored by FNB, the Office of the KwaZulu-Natal Premier and the provincial Ministry of Agriculture, the conference was preceded by a tour of significant agricultural areas in KwaZulu-Natal which culminated in a function at the Showgrounds. Its attractive anniversary tie and prospectus were deployed to take further advantage of this favourable marketing opportunity for the RAS and its activities.

The occasion helped to put the Society firmly back on the international map. This followed a long period of non-membership of the RASC between South Africa's declaration of a Republic in 1961 and its re-admission to the Commonwealth following the 1994 democratic election. Just a few months after that historic event the RAS had successfully re-applied for membership to the organisation having been invited to do so by its president, HRH The Duke of Edinburgh.

The primary objective of the RASC was the interchange of agricultural information among member societies primarily through biennial conferences. The value of membership became evident to Derek Spencer when representing the Society and the country at the RASC conference in Chester at his own expense, except for the registration fees which the RAS paid.

David Fowler submitted a paper there on the future of agricultural shows and represented the Society at the subsequent gathering in Darwin. So too did Derek Spencer who took great pleasure in attending several more conferences in other parts of the world which proved to be very informative, not least during their regional pre-and post-conference tours.

The circumstances of the RAS had strengthened to the extent that in 1998 it was able to join other member societies throughout the world in making a R2 000 contribution to the RASC head office in London to ease its weak financial position. In response to an invitation from Buckingham Palace, in July 2000 John and Fiona Fowler represented the Society at a royal garden party.[58]

As May 2001 approached the RAS executive was very conscious of the successful earlier anniversary Shows which had been held in 1953 (centenary), 1976 (125th), 1981 (republic festival), 1984 (125th Show) and 1988 (Pieter-maritzburg's 150th). Despite the threat posed by the recent outbreak of foot and mouth disease in the region, the Royal Show which followed in this the Society's 150th anniversary year was described as 'one of the finest in recent times'.

It comprised a 'full and varied entertainment programme', including a good turnout of stud cattle, as well as a spectacular fireworks display in consultation with the SPCA. While the Ukhozi FM Festival helped to attract the second largest overall attendance in its history, the Society's 150th anniversary tie left no doubt that agriculture was still its primary emphasis.[59]

In 2001 the RAS could claim to be in sound condition with 150 years of experience and 142 Shows behind it, an established position in the community, secure finances and competent managerial staff at its service. The Society's competitive sections and other activities had enjoyed and, in some cases, continued to enjoy the support of numerous volunteers. It had also developed close and effective relationships with several reliable contractors, as well as caterers and other service providers.[60]

In keeping with institutional practice at that time, by the turn of the century the RAS had formulated a mission statement. Following strategy meetings to discuss its content, it was finalised by November 1997 and first published as part of the annual report in 1999. This clearly defined its nature and objectives, acknowledging that it had long since expanded its initial focus on matters agricultural to embrace commerce and industry as well:

The Royal Agricultural Society of Natal is a non-profit making organisation committed to serving the interests of Agriculture, Commerce and Industry in the Province of KwaZulu-Natal in particular and South Africa as a whole. Accordingly, we aspire to facilitate interaction, education and trade through the medium of exhibitions and shows. In furthering this objective and fully appreciating that financial viability and the optimal utilisation of all resources cannot be compromised, we are ever mindful of

- our partnership with the City and Region of Pietermaritzburg-Msunduzi and the Province of KwaZulu-Natal;
- our social responsibility to the wider community;
- our heritage and traditions.

In fulfilling this mission, we subscribe to the principles of 'fair play', transparency and openness in all our dealings.[61]

The Society's compass had been firmly set for the new millennium.

ENDNOTES

1 This chapter makes extensive use of *AR*s 1995–2002.
2 RASM, Horse, 26 July 1994: 5; Marketing, 3 November 1994: 1–2, 27 January 1995: 1, 10 March 1995: 2, 9 February 1996: 2-4, 28 November 1997: 1 and 8 April 1998: 1; Executive 25 January 1996: 2. Shute, 'Royal Show': 34, 37.
3 *AR* 1998: 5. Shute, 'Royal Show': 40–41. RASM, Commerce and Industries, 26 August 1998: 1; Executive, 16 September 1998: 2, Terry Strachan, interview, 6 March 2019.
4 Derek Spencer, interview, 29 May 2019.
5 *AR* 2000: 5. RASM, Horse, 11 August 1997: 4.
6 RASM, Grounds and Maintenance, 30 June 1995: 2, 12 January 1996: 1, 14 June 1996: 1–3 and 12 August 1996: 1–2; Horse, 11 October 1995: 3. Bert Cornell, interview, 18 June 2019.
7 RASM, Management, 7 November 1995: 5 and 27 February 1996: 1; Executive, 27 June 2000: 5.
8 Jenny Fraser, e-mail, 1 July 2019.
9 *AR* 2000: 5. Di Fitzsimons, interview, 4 June 2019.
10 Bert Cornell, interview, 18 June 2019. Reeva Wing, interview, 5 July 2019. Jenny Fraser, e-mail, 1 July 2019.
11 Reeva Wing, interview, 5 July 2019.
12 Shute, 'Royal Show': 53. RASM, Horse, 13 October 1998: 1–2; Management, 18 November 1998: 2 and 20 January 1999: 1; AGM 25 October 2000: 4. Terry Strachan, interview, 6 March 2019.
13 Bert Cornell, interview, 18 June 2019.
14 *AR* 1995: 7. RASM, Beef Sub-Committee, 20 September 1994: 1; Commercial Cattle Sub-Committee, 7 November 1994: 2.
15 Moyra Poole, interview, 1 August 2019.
16 Iona Stewart Personal Papers, 'Beginning of it all'. Iona Stewart, interview, 29 October 2019. John Fowler, interview, 10 June 2019. RASM, Dual Purpose Sub-Committee, 26 October 1995: 3; Cattle, 2 July 1996: 3 and 23 July 1997: 2; Beef Cattle Sub-Committee, 12 September 1996: 2; Marketing, 15 November 1996: 3; Grounds and Maintenance, 24 July 1998: 2.
17 RASM, Cattle, 10 March 1999: 2 and 29 June 1999: 2–3.

18 Shute, 'Royal Show': 56.
19 RASM, Sheep and Goat, 4 October 2000: 2.
20 RASM, Pigs, 29 March 1995: 1–2; Pig Think Tank, 12 July 1995: 1–2. Mike Moncur, interview, 26 June 2019. Shute, 'Royal Show': 38.
21 RASM, Apiarian, 1 November 1995: 2.
22 RASM, Agriculture and Produce, 22 February 1995: 1; Combined Chairmen, 19 January 1999: 1; Management, 18 January 2000: 1; Sheep and Goat, 10 March 2000: 1; Agricultural Implements, 4 August 2000: 2. Shute, 'Royal Show': 53–54.
23 RASM, Executive, 26 June 2001: 4; Management, 17 July 2001: 1, 21 August 2001: 1; General Manager's Report, 21 August 2001: 2.
24 Iona Stewart, interview, 12 February 2019. Derek Spencer, interview, 29 May 2019. Di Fitzsimons, interview, 4 June 2019. RASM, Management, 17 September 1997: 2. Jeanne Mather, 'Crafts Hall 2000: an impression' (unpublished typescript, 2000 in R. McDonald Miscellaneous Correspondence).
25 AR 2001: 9. RASM, Commerce and Industries, 16 August 1996: 3. Shute, 'Royal Show': 34, 36, 38.
26 RASM, Commerce and Industries, 17 May 1996: 2. Iona Stewart Personal Papers, 'Beginning of it all'.
27 AR 2001: 9.
28 AR 1999: 8. Shute, 'Royal Show': 53.
29 AR 2001: 9. Iona Stewart Personal Papers, 'Beginning of it all'. Shute, 'Royal Show': 55.
30 AR 1996: 10.
31 RASM, Finance and Capital Projects, 17 February 1995: 2.
32 Derek Spencer Personal Papers, 'Opening address at the 1998 Royal Show, May 1998': 5.
33 RASM, Executive, 29 November 1994: 2–3, 25 June 1996: 1 and 7 August 1996: 2; General, 4 August 1995: 3; Special General Meeting, 18 October 1995: 2; Management, 23 July 1996: 2. Shute. 'Royal Show': 35–36.
34 Jenny Fraser, e-mail, 1 July 2019.
35 Terry Strachan, interviews, 6 and 12 March 2019.
36 Iona Stewart Personal Papers, 'Beginning of it all'. Shute, 'Royal Show': 37. RASM, Executive, 13 November 1996: 2 and 1 July 1997: 4; Poultry and Bird Clubs, 5 April 1997: 2; Sheep and Goat, 16 July 1997: 2; Finance, 12 February 1998: 1; Marketing, 8 April 1998: 1. Terry Strachan, interviews, 6 and 12 March 2019. Derek Spencer, interview, 29 May 2019. Ron McDonald, interview, 12 June 2019.
37 Shute, 'Royal Show': 43–44, 52. RASM, Grounds and Maintenance, 7 March 1997: 1; AGM, 21 October 1998: 2.
38 RASM, Commerce and Industries, 6 February 1998: 3; Grounds and Maintenance, 6 March 1998: 1.
39 RASM, Executive, 15 November 1995: 2; Garden Show General, 22 October 1996: 2.
40 RASM, Garden and Leisure Show Ideas, 6 November 1997: 3; Marketing, 28 November 1997: 1; Management, 13 October 1999: 12; Executive, 20 October 1999: 4; Garden Show, 31 March 2000: 2. Shute, 'Royal Show': 48, 54.
41 RASM, General, 14 October 2000: 2.
42 RASM, Executive, 12 October 1994: 2 and 7 March 1995: 3; Commerce and Industries, 14 October 1994: 3; Management, 25 January 1996: 1–2 and 21 May 1996: 1. Shute, 'Royal Show': 37.
43 Terry Strachan, interview, 12 March 2019. RASM, Finance, 23 October 1995: 1; Management, 27 February 1996: 2; AGM, 22 October 1997: 4.
44 RASM, Apiarian Section, 22 January 1998: 2; Marketing, 6 December 1999: 1.

45 RASM, Agriculture and Produce, 2 August 1995: 4; Management, 15 October 1997: 1, 18 November 1998: 1, 13 October 1999: 2 and 19 June 2001: 2; AGM, 20 October 1999: 2; Executive, 20 October 1999: 2–3, Commerce and Industries, 13 October 2000: 1. Shute, 'Royal Show': 53–54. Terry Strachan, interview, 12 March 2019.

46 *AR* 2001: 5. Shute, 'Royal Show': 56.

47 RASM, AGM, 20 October 1999: 3.

48 RASM, AGM, 12 October 1994: 3; Executive, 29 August 1995: 2–3.

49 Margaret von Klemperer, e-mail, 6 July 2019. Bert Cornell, interview, 18 June 2019. Tony Hesp, 'Hubert Ralph von Klemperer' *Natalia* 29 (1999): 102–105.

50 RASM, Executive, 7 May 1997: 2, 18 September 1997: 5, 16 September 1998: 6, 3 February 1999: 4, 28 April 1999: 4, 25 April 2000: 4, 19 September 2000: 4 and 24 October 2001: 2; General Manager's Report, 14 March 2000: 2. Shute, 'Royal Show': 35. Cynthia Dreboldt, interview, 5 March 2018.

51 RASM, Commerce and Industries, 30 August 1995: 2 and 17 May 1996: 3–4.

52 RASM, Management, 25 January 1996: 2 and 27 February 1996: 1; Executive, 25 January 1996, 'Development proposal', 19 March 1996: 1–2, 'President's opening remarks, 18 September 1997: 3 and JHI Property Services report, 16 September 1997, 5 November 1997: 2 and 7 March 2000: 2; Grounds and Maintenance, 7 March 1997: 3.

53 RASM, Grounds and Maintenance, 12 January 1996: 2, 8 March 1996: 1–2 and 7 November 1997: 1; Executive, 23 April 1997: 2; Management, 17 June 1997: 2 and 20 August 1997: 2.

54 RASM, Finance, 16 April 1998: 2; Management, 15 September 1998: 3 and 18 November 1998: 2; Executive, 16 September 1998: 3 and 21 October 1998: 2.

55 *AR* 2001: 11. RASM, Management, 18 July 2000: 3 and 21 November 2000: 1; Garden and Leisure Show, 30 March 2001: 3; Executive, 28 August 2001: 3 and 28 August 2001: 3. Shute, 'Royal Show': 52, 56, 59.

56 Reeva Wing, interview, 5 July 2019.

57 ibid.

58 Iona Stewart Personal Papers, 'Beginning of it all'. Derek Spencer, interview, 29 May 2019. RASM, Executive, 30 June 1994: 10, 19 September 2000: 4 and 27 January 2009: 6; Management, 19 August 1998: 3. John Fowler, interview, 10 June 2019. Shute, 'Royal Show': 40, 42–43, 60.

59 *AR* 2001: 5. RASM, AGM, 24 October 2001: 4. Shute. 'Royal Show': 57–59.

60 Shute, 'Royal Show': 50–51, 60.

61 *AR* 1999: inside front cover. RASM, Executive, 18 September 1997: 2; Management, 19 November 1997: 3.

3   'UPHOLDING THE TENETS OF EXCELLENCE', 2002–2011

THE PERIOD BETWEEN the 150th anniversary of the RAS in 2001 and its 160th in 2011 was no different to previous decades in presenting various challenges and opportunities, some familiar and others entirely new. These included a major economic downturn, which impacted severely on the exhibition industry worldwide resulting in the closure of numerous shows, among them the parent Royal Show in England in 2009.[1]

That year the 150th Royal Show was held in Pietermaritzburg, eight years after the Society's 150th anniversary. Unlike the 125th in 1984, it passed largely unnoticed in the prevailing 'challenging conditions'. There was also concern about the deteriorating situation of the farming community in Zimbabwe with which KwaZulu-Natal had 'always enjoyed a close affinity' and there were mounting security issues in the local agricultural sector.[2]

In these unfavourable circumstances, coupled with disconcerting socio-political uncertainties in South Africa, a professional Matrix assessment was commissioned to ascertain what contribution the RAS actually made to the community at large. It concluded that the Society's various activities generated and maintained more than 2 000 jobs in the Natal Midlands area and injected in excess of R150 million into the regional economy. It was also successfully maintaining its reputation in the 'international fraternity' of exhibitions as an organisation that was committed, as its president Garth Ellis put it, to 'Upholding the Tenets of Excellence'.[3]

The RAS continued to owe much to sound management and careful pre-emptive planning, to a small but dedicated team of permanent employees, and to a committed core of active volunteer supporters. Success was also due to a careful mix of attractions at its annual Royal Show and to hosting an increasing variety of income-generating activities.

It still never lost sight of its agricultural roots, even though this dimension ran at a net loss in purely financial terms. CEO Terry Strachan described it as 'a balancing act' involving numerous often competing components some of which, like popular concerts, made cross-subsidisation of the agricultural

G.I. (Garth) Ellis grew up in Pietermaritzburg and was a member of the RAS from childhood. In 1985 he became a pedigree beef breeder at Winfield Farm, Seven Oaks promoting his Santa Gertrudis livestock at the Royal Show and others. He was a breed representative on the cattle committee and also served as a steward in that section. After being invited onto the executive from 1999 as one of a group of new younger members, he served a term as vice-president (2004–2006) before becoming president (2007–2009). It was an interesting though sometimes difficult experience, the most controversial incident being the decision (not his own) to include a discreet sex shop at one of the Shows. It was not exactly part of the RAS tradition and was not repeated. Ellis subsequently became a committee member and trustee of the Royal Mews Trust committee and of a three-person advisory committee on possible future scenarios for the RAS.[4]

sections possible.[5]

The **Main Arena** displays at the yearly Show provided ample evidence of what did and did not attract the paying public. By 2002 it was clear that, apart from the concerts, major Main Arena events were not as well patronised as in previous years. In contrast, micro entertainment like the Food Hall, sheep shearing demonstrations, dog shows and River Stage performances were gaining popularity.[6]

Expectations that Prince Andrew, the Duke of York, would open the 2002 Show were disappointed but other luminaries were persuaded to oblige in subsequent years. These included KwaZulu-Natal Premier Dr Lionel Mtshali in 2003 when, unusually, the opening ceremony took place in the cattle arena. Several provincial cabinet ministers visited the Shows including a subsequent premier, Sbu Ndebele. In 2005 the guest of honour was the Hon. Mr Justice V.E.M. Tshabalala, Judge-President of KwaZulu-Natal.

For three years the opening ceremony took the form of a cocktail party until 2007 when, for the first time in the Society's history, there was no such event at all. This was due to the numerous other functions taking place each year in connection with specific aspects of the Show. Hardly anybody seemed to notice the absence of an opening ceremony and no objections were raised. In 2009 the entire provincial cabinet was present one afternoon while the MEC for Finance, Ina Cronjé, and Msunduzi's municipal manager, Rob Haswell, attended the corporate sponsors' dinner.[7]

In 2003 the SANDF made a welcome return to the Main Arena and as an exhibitor. Heavy summer rains waterlogged that venue in 2005 causing serious damage, but the following year was particularly memorable for the popular joint performance of the South African Lipizzaners and the Drakensberg Boys' Choir.

In 2007 the Dundee Diehards re-enactment of the 1879 battles of Isandlwana and Rorke's Drift and the SAAF's Rooivalk helicopter display provided the highlights. The Lipizzaners returned in 2008 to public acclaim, along with the ever-popular Soweto String Quartet while American Dave Smith, the human cannonball, provided a further explosive attraction.

Four big musical events were held in 2009, including the Royal Folk Show and a rock extravaganza while Radio Hindvani hosted the Royal Bollywood extravaganza for the Indian community and Afrikaans-speakers welcomed the return of an RSG concert. Other Main Arena attractions included a freestyle motorcycle demonstration, the SAPS task team, the national dog agility championships, sheep dog herding and the Wheel of Death. The Ukhozi concert, which for years had been held as a concluding event at the Show drawing about 30 000 devotees, attracted only 12 214 as a separate event.

It was successfully reincorporated on the last Sunday the following year with 21 000 fans packing the Main Arena. The Royal rock concert and the Royal Bollywood extravaganza were also successfully held during the final weekend although there was growing concern that visitors on those days were being subjected to demographically-orientated concerts with little or no agricultural livestock still to be seen.

Fortunately, there was sufficient time between Shows to repair the damage that had been caused to the grounds and some of its facilities by high winds and rain in August 2009. In 2010 the Lipizzaner display was again a crowd favourite. That year the SANDF performed two full military tattoos on the first two evenings, which was a first for Pietermaritzburg. It was also the first time that the cost of providing these additional entertainments exceeded R1 million.

More military tattoos followed in 2011, together with a Bobby Angel/Dennis East/Billy Forrest musical nostalgia trip, a Johnny Clegg concert, the now usual Royal Bollywood show and an orchestral concert. Not least, the Ukhozi FM bash attracted a 25 000-strong audience being sponsored by what had purportedly become the world's second largest radio station.

There was also a fireworks display to celebrate the Society's 160th anniversary. Some years previously, before such expensive events had been dropped from the regular programme and confined to special occasions, there had been public concern about the stress they caused to animals in the Showgrounds and to domestic pets in the neighbourhood.

In response RAS executive member and animal scientist Iona Stewart had visited the cattle stalls during a fireworks display. She was 'pleasantly surprised to see how calm they were' while some of the Brahmans, 'possibly the most highly strung of all the breeds, raised their heads, looked in the direction of bangs and lights, and then went on feeding from their troughs'.

A survey of local residents within a 4-kilometre radius established that 79% 'were not opposed to a well communicated fireworks display' held at an advertised 'given time'. The SPCA agreed, although it did not favour such events. The expense involved in these displays ensured that, in future, they would indeed only be held on special occasions. By then there were complaints about the noise emanating from the funfair and from the new chainsaw racing competition which subsequently featured on sports channel television.[8]

The **Agricultural sections**, not to be outdone by all these distractions, still formed the core of the Royal Show. This was revitalised by the Society's ongoing conscious commitment to return to its agricultural roots and to upgrade that vital part of the Show from provincial to national level.

The boys of Weston and Zakhe colleges assisted in performing numerous useful tasks in exchange for being educated about the animal breeds they were handling. Popular innovations such as a spit braai for farmers and a grooms' tug-of-war were continued to attract more exhibitors.

In addition, efforts were made to encourage Natal Agri, the largest co-op in the province, various fertiliser distributors as well as the coastal sugar farming community, to participate more actively in the Show. In 2002 invitations and entry tickets were distributed to farmers associations via Kwanalu. An agricultural leaflet was sent to all implement companies and was printed in the *Stock Owners/Kwanalu News* and *Farmer's Weekly*.

Terry Strachan had always made it clear that he was resolved to put the 'A' back into RAS and promoted the agricultural sections, financially and otherwise, as best he could. At a time when agricultural societies elsewhere in the country were struggling to attract exhibitors the Royal Show gained noticeably more attention from farmers outside the province. While few of them may have been royalists, Strachan was also at pains to emphasise the 'R' in RAS because the royal assent to call itself the Royal Agricultural Society was another aspect of its proud tradition that should not be forgotten.[9]

In 2003 the livestock programmes were extended by two days from Sunday through to Friday at the request of the farming community and despite the cost and time implications. It resulted in the highest participation in these sections in fifteen years with nearly 900 cattle and 500 sheep being exhibited.

By 2004 all nine provinces were well represented. By 2009 it was reckoned that approximately 40% of participants in the agricultural sections were from elsewhere in South Africa and that Afrikaans was as frequently heard in the Show rings as English. Further, that in recent years there had been a 138% increase in the number of farmers attending, a 76% rise in cattle entries and a 520% increase in the Sheep Section. Agricultural implement exhibits had risen by 78%.

The **Cattle Section** witnessed the participation of several first time Angus, Brahman, Hereford and Limousin exhibitors in 2002, with a noticeable increase in Brahmans and Jerseys. An invitation to the Zulu King to exhibit some of his nGuni cattle, with transport provided, was not taken up.[10] The following year, in addition to the extraordinarily high number of entries, the construction of two floodlight towers adjacent to the cattle arena was immediately followed by a new nocturnal exhibition of Holsteins. This extended livestock viewing was very popular with the public.

In 2004 all breeds were represented with the 134 Brahmans dominant. The following year the Royal Show was chosen to host the Standard Bank Jersey Expo, widely regarded as 'the most prestigious dairy event ever held in South Africa'.[11] Some of the 227 entries came from as far afield as the Western Cape and, not to be outdone, the Simmentaler Club held a highly successful sale that year.

In 2006 the Show hosted the Ayshire, Brahman and Holstein national championships with every cattle stall in use for the first time in several years. The Holstein Club erected a massive marquee, similar to those at royal shows elsewhere in the Commonwealth, to accommodate exhibits, judging and an entertainment programme. The national president of the Brahman Breed Society described it as the best show he had attended in 30 years.

The following year the cattle programme was extended by a further day so that the public could view all 728 entries. These were of a high quality despite the prevailing drought and for the first time the prizewinners were announced at a Standard Bank-sponsored Gold Cup dinner. The 2008 Show featured eleven cattle breeds from 53 exhibitors with 476 entries in the showing classes and 118 commercial animals. The Future Farmers classes were well supported, primarily by those entering the students' challenge in the Commercial Section and were boosted by the first-time participation of the Owen Sithole College of Agriculture in Empangeni.

As usual in 2009 Brahmans were numerically the most prominent among the 740 entries but they also included 30 Pinzgauers from the Tzaneen region, a breed that had not competed at the Royal Show for many years. At Judy Stuart's suggestion a competition for non-stud commercial heifer classes was also re-introduced to enable farmers to view and buy good quality in-calf animals that did not necessarily achieve stud grade standards.

The 2010 Show attracted entries from all over South Africa and from Namibia. Among other breeds Braunvieh, Ayrshire and the SA Huguenot made a welcome re-appearance after long absences while Doug Savage (Canada) and James Playfair-Hannay (Britain) served as judges. An exhibition of Boran, Bonsmara and Brangus took place concurrently with the Show and a new Pinzgauer-nGuni crossbreed named PinZ²yl was shown.

Despite an apparent foot and mouth outbreak in northern KwaZulu-Natal there were 945 cattle entries at the 2011 Show, 55% of them from outside the province. The Hereford national championships that year attracted nineteen exhibitors and 157 animals. Jan Wills, secretary-general of the Hereford World Council, attended and animal nutrition specialist Dr Cliff Lamb from the University of Florida was the senior judge. He also addressed the Hereford Breed Society's symposium on fertility, profitability and tick management.

There were no dominant exhibitors among the major award winners during this period. Several were from outside the province which was hardly surprising considering the increasing numbers of entries attracted from all over the country. John Devonport of Devlan Farm won the RAS Gold Cup for the beef bull supreme champion in 2005 and the Sutherland trophy for the beef bull reserve champion in 2006, both Limousins.

K.A. Haug won the latter title in 2007 and the supreme championship in 2008, both with Brahman Reds. Silence Genetics won the supreme championship with a Pinzgauer in 2009 and Phillip de Waal from the Western Cape did so in 2011 with a Hereford. White Horn Santa Ranch won the Meadow Feed Trophy for the beef cow reserve champion in 2002 and the RAS Gold Cup for the beef cow supreme champion in 2003, both with Santa Gertrudis entries. John Devonport was also prominent in this category winning the supreme championship in 2004 and 2007 and the reserve trophy in 2006, again with Limousin specimens.

There were no noticeably successful competitors for the RAS Gold Cup and John Simpson Memorial Trophy awarded to the dairy supreme and reserve champions each year although Jerseys and Holsteins were the usual winners. Jonsson's Jerseys won the supreme title in 2002 and the reserve championship in 2011 when Free Stater Wouter du Plooy from Bultfontein exhibited the best dairy cow, a Holstein. Apart from the bovine stud classes, commercial heifers competed in a separate competition for the third successive year and attracted more than 100 entries.

The **Pig Section** had long-since ceased to exist as a separate entity at the Royal Show but the **Sheep Section** experienced a doubling of entries to 234 in 2002 with exhibitors and visitors showing increasing interest. This was due partly to Kwanalu, which attracted its members' attention to the Show, as well as to the Hampshire Down championship held there that year. The RAS executive nevertheless acknowledged Mike Moncur's enthusiastic efforts in promoting a significant revival of the Sheep Section.

The hand and machine sheep shearing displays continued to be popular and for the first time the Damara breed was exhibited. Despite ongoing concern about the increasing expense involved in exhibiting livestock in 2003 the Ile de France national championships were held at the Show. The facilities and hospitality provided were sufficient to attract both the Damara and Hampshire Down championships the following year with the former being moved for the first time from Bloemfontein.

That year, with Marlize du Preez's input, there was also a prize-winning expo of wool, fibre and mutton products. In addition to the now traditional spit braai a sheep dinner was held in conjunction with FNB in 2003, primarily to welcome the Ile de France exhibitors. In subsequent years FNB and the *Farmer's Weekly,* among others, continued to co-host a number of similar social functions.

In 2005 the wool and mohair expo included a daily fashion show while Suffolks made a re-appearance for the first time in ten years. The following year the Show hosted the Hampshire Down and Suffolk national championships in addition to the three competitions in the Cattle Section. The 600 sheep entries in 2007, which included another Hampshire Down national championship and a large number of Merinos, stretched the available facilities for the first time in several years but successfully included an expo which offered a one-day training session for emerging farmers.

The Sheep Section continued to develop under Mike Moncur's leadership while Marlize du Preez maintained an interesting variety at the gold medal-winning expo. This included commercial displays, weaving and sheep shearing with Harry Tully and P. Eustace providing the necessary animals. The involvement of youngsters entering sheep and goats had become firmly entrenched. In 2008 indigenous goats were introduced for the first time with the intention, assisted by the provincial Department of Agriculture, of improving the knowledge and output of emerging farmers.

The 540 sheep entries in 2009 included competitors in the Hampshire Down and Suffolk national championships as well as increasing numbers of Merinos and another indigenous goat component with the emphasis again on emerging farmers. The sheep, wool and goat expo included the usual sheep shearing as well as an exhibition of indigenous small stock and retail sales ranging from cheese to woollen socks. Twee Seuns Suffolks won both the supreme champion ewe and champion lamb on the hoof trophies.

In 2010 there were more than 2 000 animals on show altogether with entrants from as far afield as Namibia. Most sheep breeds were represented with the Hampshire Down and Suffolk national championships being held yet again. The sheep and goat expo continued to flourish under Marlize du Preez's direction, ably assisted by Jenny Bester. Both the Kalahari Red Goat and SA Boer Goat displays attracted particular attention while the return of a fashion show provided additional interest. When this acknowledged 'show within a show' enjoyed its 12th year in 2011 it could claim to be representative of 'the entire supply chain involved in natural fibres (and associated products)'.[12]

The Youth Show and Future Farmers competitions were actively encouraged with more than 100 youngsters ranging in age from four to eighteen participating, including pupils from the Weston, Vryheid and Zakhe agricultural colleges. For the first time the breeders' societies were assisted with an expo focused on branding and marketing, in line with what had taken place at shows elsewhere in the Commonwealth.

The most prominent winner of the two prizes offered in the Sheep Section was R.J. (Russell) Shorten of Aliwal North. Shorten & Son won the William Cooper & Nephews Floating Trophy for the supreme champion ram in 2004 and 2006 and the Epol Floating Trophy for the supreme champion ewe from 2004 to 2008 inclusive, all with Hampshire Down specimens.

Rodney Dorning won the Epol Trophy in 2002 and 2003 with Hampshire Downs while Twee Seuns won the ram championship in 2008 and the ewe title in 2009 with Suffolks. These were the most successful breeds exhibited, although Ile de France specimens won both the supreme awards in 2011, with T.A. Clarke from the Cape exhibiting the champion ram and Henk Moller of Amersfoort the champion ewe. Albert van Zyl scooped almost all the awards in the goat classes.

The **Cattle and Lamb Carcass competitions** meanwhile were combined from 2002, through the efforts of Mike Moncur, Rodney Dorning and others, consolidating them as the major national red meat event of the year.[13] The commercial dimension of both the Cattle and Sheep sections enjoyed the support of Ben Booysen, the Red Meat Producers Organisation (RPO), Stock Owners, Voermol and Crafcor Hygrade abattoir whose closure in 2008 constituted a loss of important sponsorship.

When in 2003 K. Forster's champion lamb carcass sold for R315 per kg it was said to be a South African record. The R300 per kg which Russell Shorten secured for his Hampshire cross in 2005 was second only to that price while his reserve mutton champion carcass, also a Hampshire cross, sold at R210 per kg. John and Peter Mapstone, who had produced the champion beef carcass in 2003, also scored a double in 2005, producing the champion beef carcass, a Sussex, which fetched R26 per kg and the reserve champion which sold at R20 per kg.

Records were broken again in 2006 when G.J.L. (Gert) Lotter of the Eastern Cape won the lamb carcass competition with a purebred Dorper that scored a 95.81% rating and sold for R13 000 or R600 per kg to Shaam Rajpul's Butchery in Durban. Regular buyer Jeremy Shearer of the Butchery Restaurant in Pietermaritzburg paid a record South African price of R18 000 or R71 per kg for the champion beef carcass.

The following year the 21.5 kg champion lamb carcass broke the national record yet again by selling for R64 500 or R3 000 per kg which was also believed to be a world record. Mike Moncur recalled that, as the auction price climbed to that level, he assured Garth Ellis that the competing bidders were well aware it was per kg and not for the whole carcass![14] Mark Gietzmann and Jimmy Keyser of DeBonnaire Stud, Fochville bred the Ile de France animal in question which was rated at 98.15%. Ivan Rattham of Durban's Crown Butchery bought it for marketing purposes, as was often the case, because of the prestige associated with the Show's winning carcasses.

C.A. Froneman of Biggarsgat Brahmans, Dundee produced the champion 238.5 kg beef carcass, rated at 97.35%, for which Jeremy Shearer paid R14 310 or R60 per kg, well above the current R19.50 ruling price. Following the closure of the Cato Ridge Abattoir arrangements were made with Jacques Abattoir in Estcourt (for sheep) and Triple A Abattoir in Cramond (for beef) making it possible to continue the carcass competition in 2008.

That year the Society did well in the South African Meat Industries Company (SAMIC) red meat competition which reflected favourably on the quality of breeding and feeding maintained by the region's farmers. SAMIC, which oversaw the technical standards relevant to the country's red meat industry, subsequently declared the Royal Show to be the best agricultural show in South Africa, as did the Land Bank.

The champion beef carcass in 2008, a Hereford, enjoyed the highest grading ever attained at the Show, a near-perfect 99.97%. It was a great achievement for the Cedara Agricultural College where the animal was bred and although the R38 per kg price which it fetched was disappointing this was 80% above the ruling market price. The reserve champion carcass, a Bonsmara, was not far behind scoring 99.5% and selling at R32 per kg. Participation in the carcass competitions was a valuable learning experience for agricultural college students and for members of the local university's Faculty of Agriculture.[15]

The winning lamb carcass and the first and second reserve champions were all white Dorpers which the Lotter family had bred. Remarkably, they achieved 99.15%, 97.96% and 97.82% respectively with the winning carcass selling for R200 per kg or six times the ruling price.

Dorpers swept the boards again in 2009. C. Reitz of Williston, Northern Cape achieved 98.98% with a carcass that sold for R42 per kg compared with the R32 ruling price. Cedara Agricultural College again produced the champion beef carcass, a Bonsmara which scored 98.67% and was sold to Jeremy Shearer at R43 per kg compared to the ruling price of R23.50. J.A. Labuschagne of Dundee won the first reserve and champion beef group with his Brahmans.

In 2010, against over 100 competitors, Gert Lotter again entered the champion, a Dorper, which achieved a 99.62% score and sold for R600 per kg! His champion group, also Dorpers, sold for R100 per kg when the ruling price was R38. The following year, although the scores and sale prices were not as high, Lotter scooped the prizes yet again.

There were 146 beef entries in 2010 with Weston Agricultural College producing the champion carcass, a Braford, which scored 98.11% and sold for R50 per kg. Weston won all the prizes in 2011 with Brahmans (champion, reserve champion, champion group and reserve champion group) with the top specimen scoring 98.88% and selling at R65 per kg.

In the Cattle Section the title 'champion commercial animal on the hoof' was fairly evenly distributed at this time with Garth Ellis winning it in 2004, 2006 and 2009 with Santa Gertrudis specimens. The most successful exhibitor in the champion lamb on the hoof category was, again, Shorten & Son who won in 2004, 2005 and 2006 with Hampshire Down entries while the Lotter family won in 2007 and 2008 with Dorpers.

Other **Agricultural sections** at the Royal Show included the Youth Show and Future Farmers competitions in an ongoing effort to widen the age range of participants. **The Apiarian, Bird and Poultry and Rabbit sections**, as before, broadened the spectrum of interest in the Show and attracted public attention with informative displays intended as much to educate as to entertain.

The **Apiarian Section** enjoyed renewed participation after a previous decline and seemed well on its way to becoming the preferred national venue for such shows. Traditionally it also regularly participated in the Garden Show with presentations of mead (honey wine), confectionary, items made from beeswax and tools used in the apiarian industry.

The 2004 South African Bee Congress held at the Royal Show attracted 125 delegates. The following year as the section grew stronger it doubled its sales while providing an interactive learning experience for interested members of the public at both the Royal and the Garden shows. In 2007 Justin Herd won seven of the nine prizes, William Urquhart produced the best bottled honey and Delville Stephenson was awarded the novice floating trophy.

Urquhart won again in 2008 but entries were affected by the late flowering of eucalyptus trees which resulted in the province's honey production being six weeks behind schedule. The Honey Hall won a gold medal for excellence, thanks to the KwaZulu-Natal Bee Farmers Association's refurbishment of the venue and its interesting interactive displays. These included photographs and posters illustrating the life cycle of bees as well as an observation hive which attracted much attention. The Honey Hall also won gold at the Garden Show.

In 2009 Ron Botha of Highflats Apiaries produced the champion bottle of honey with a score of 99.5%. It contained some highly sought-after Hlalwane (*Isoglossa woodii*), a plant that reputedly only flowers every seven to ten years in KwaZulu-Natal's river valleys. The Bee Farmers Association organised a display which explained the highly dangerous American foulbrood disease that could cause heavy bee losses and had recently been detected in the Western Cape but had hopefully been contained by effective quarantining.

In 2010, the centenary of formal beekeeping in the province, the Honey Hall again won a gold medallion. The Show saw a pleasing increase in entries as well as the participation of Caroline Whitehead from Gauteng with her variety of products and Caroline's VW Caddy which provided illustrations of bees and honey combs. The marketing initiative launched that year produced dividends in 2011 with several new entries from elsewhere in South Africa and a further sales increase. Craig Campbell succeeded William Urquhart as chairman of the KwaZulu-Natal Bee Farmers Association.

The **Bird and Poultry Section** remained, as Terry Strachan was at pains to stress, 'an integral part of the soul of the Society' which would always be assisted 'where possible'.[16] It attracted more than 1 800 exhibits in 2004 and increasingly used the Showgrounds for several competitions at other times of the year.

In 2005 the RAS hosted the national Poultry Championships and the biennial Pig and Poultry Expo for the first time. As interest in purebred poultry increased record sales were achieved that year as well as enhancing the educational dimension. Budgerigar sales also rose and the 485 entries in 2007 were drawn from all over the country. By then that section was clearly beginning to outgrow the two halls allocated to it.

All parts of it enjoyed a successful Show in 2008 with the pigeon category attracting 515 entries representing 76 different breeds, some of them new to the country. Justin Dent won the junior class with an Old German Owl and Gary Shultz the senior class with a Giant Pouter. The Natal and Coast Poultry Club attracted more than 600 entries as well as significantly more exhibitors.

The following year the fur and feather committee, representing both bird and rabbit exhibitors, was established to give them a collective voice.[17] The South African Fancy Pigeon Association displayed 400 specimens during a full seven-day exhibition while the local Budgerigar Club sold 60 birds. The Canary and Cage Bird Club and Natal and Coast Poultry Club attracted the usual attention.

In 2010 the latter celebrated its 125th anniversary with nearly 600 birds exhibited in addition to its activities outside Royal Show time. By 2011, when it could claim to be the third largest poultry club in South Africa with 48 members, it featured on Agri TV and hosted Christopher Parker, president of the Poultry Club of Great Britain as a judge. That year the Pietermaritzburg Canary and Cage Bird Club also attracted publicity when it celebrated its 99th anniversary and in all there were 800 birds on show.

The **Rabbit Section** similarly went from strength to strength, experiencing a record number of entries in 2004 with spinning demonstrations and a display of Angora products providing additional interest. The following year some exhibits had to be removed immediately after judging due to a space shortage. By 2006 there were as many as 300 exhibits with increasing numbers of rare breeds generating keener competition. By the following year demand for commercial varieties had grown noticeably as the sale of rabbits and rabbit products increased.

There was another good year in 2008 with increasing interest from emerging farmers and 378 entries necessitated the use of portable cages to cope with the overflow. In 2010 an additional 50 cages were built to accommodate the rising numbers, especially from Gauteng, with every breed existing in South Africa being represented. Public interest and the sale of rabbits and rabbit products continued to increase in 2011.

Agricultural machinery manufacturers were also increasingly attracted to the Show. By 2003 virtually all of them were represented, complementing the high number of livestock entries that year and encouraged by the growing numbers of farmers from all over the country who attended.

Agricultural services subsequently also participated, including National Chicks, Natal Agri and Omnia. In 2004 it was necessary to encroach upon part of the cattle arena to accommodate the overflow of agricultural implement exhibitors. All the non-livestock exhibitors considered their participation to have been very worthwhile, auguring well for the future.

The following year marked seven in a row in which virtually all the implement and tractor manufacturers were represented. This was due, in part, to the sterling efforts of Dave Botha and his agricultural implements committee. In 2006 construction and forestry equipment manufacturers returned to the Show after a long absence. While all exhibitors were more than satisfied with the returns from their participation one recorded no less than R8 million worth of business.

By 2007 capacity had been reached in terms of space availability and it was realised that additional areas would have to be found to avoid refusing valuable potential exhibitors. The Cattle Section obligingly made additional space available and there were still significant sales despite deepening recessionary conditions.

In July 2010 the *Farmer's Weekly* declared that the Royal Show was 'second only to Nampo' as an exhibition of agricultural equipment and tractors and that it was an ideal annual occasion for manufacturers and distributors 'to introduce new ideas and launch new equipment'. By 2011 there was a demand for more sites as existing exhibitors were requesting additional space.[18]

The **Horse Section** was listed at the end of the agricultural sections in the Society's annual reports from 1999 onwards after previously featuring much more prominently. It reflected, perhaps, a perceived decline in public interest in this once popular dimension of the Royal Show, although the driving classes were still popular.

Equitation, for South Africa's top ten junior riders, was nevertheless introduced for the first time in 2002 with sponsorship from the *Natal Witness*. KwaZulu-Natal's John Collier designed the course for junior competitions during the first weekend followed by Alan Stacey of Gauteng's construction of courses for the A-grade events. These culminated in the World Cup Qualifier which Ronnie Lawrence won on Avis Panache while Robyn Bechard's MVW Haworth was adjudged the champion show horse.

Anthony Webber from Britain judged the champion horse and show hunter classes under lights. As before, the Royal Show was fortunate to enjoy the services of several course designers including Olympic Games expert Kevin Hansen from New Zealand in 2005 and again in 2007. It also benefited from the presence at various times of international show judges including the Australians Geoff Lyall and Jan McMillan.

From 2003 equestrian events were outsourced to Showtime shows and from 2004 were referred to as the **Equestrian Section** in the Society's annual reports. In 2003, fifty C-grade horses were invited to compete at the Show and the biggest A-grade group of competitors for several years was also attracted as well as many juniors and children. Ronnie Lawrence won the South African FEI international qualifier on Avis Loutano's Son and again in 2004. Barry Taylor won that competition in 2005 and Lexi Carter the following year.

The 2003 Showtime equestrian programme was one of the most successful in recent years and included an adult ride and drive competition. The increase in entries in 2004 necessitated using both the Main and cattle arenas as well as a somewhat complicated back-to-back occupation system at the 347 stables with one horse moving in as soon as another left. With the probability of more stabling being needed in future the Society faced a challenging phase of development. After Ron Glaister, architect David Hughes and Terry Strachan had viewed the facilities at Johannesburg's Inanda Club, 50 additional stables costing R500 000 were acquired via Houses for Horses.

Gauteng riders won most of the show jumping classes that year with Christopher Hammond's Fairway's Royal Status declared the champion horse and Cape Simonsig's Oriental Express the reserve. Waterside Ready Token, bred in Balgowan, was the champion pony and the champion young rider's pony, which Elizabeth Pappas rode, was High Over's Suzie Sixpence. Junior class winners in 2005 included Chantal Chelin (JB Victor Ludorum), Ashlee Hausberger (CA champion) and Guy Houston (Young Rider class champion). The 2006 programme was similarly successful thanks in part to the efforts of Di Baxter and Charlotte Houston and despite the outbreak of African horse sickness in the interior. The following year's event was recorded as 'probably the best for many years with quality horses and top class show jumping across the board', although the shortage of adequate facilities again became evident.[19]

In the children's section Katelyn Burrows riding Africa's KitKat won the CA Victor Ludorum, while among the adult classes Jonathan Clarke on Mossandi Diadem secured the International Victor Ludorum and Sword of Honour with Shaun Neill riding Gold Rush winning the FEI world cup qualifier. Dominique Tardin rode the champion horse on show By the Book and Megan Berning the champion pony Foresyte Fatal Attraction.

Despite declining public support, the continuation of 'an equestrian presence' at the Show was still considered 'desirable'. All stakeholders were included in discussions to develop a more suitable programme for 2008 whilst 'retaining the reputation of the Showgrounds as a centre of equestrian excellence'.[20] As a result that year A-grade horse jumping was excised entirely from the Show into a separate April stand-alone event known as the Royal World Cup Qualifier. Lexi Carter won it on Pick Pocket against stiff competition from a sixteen-horse field over challenging jumps which Spanish course designer Javier Trenor Paz had prepared.

The event cost the RAS R139 000 and it was agreed that it should eventually become entirely self-funding. It attracted the biggest crowd seen at such an occasion for many years while the equestrian competitions at the Show itself assumed 'a more country-type flavour' that was rather quieter and less pressurised. They involved both adult and junior classes as well as carriage driving, tent pegging and a demonstration polo match.

The completion of 54 new stables added further value to the section and in 2009 the Main Arena was aerated and dressed with 200 tons of sifted river sand, fertiliser and mixed trace elements. This helped to improve surface standards to satisfy international criteria while the number of new stables was increased to 160.

Following the exclusion of A-grade adult jumping there was a moderate expansion of those competitions which remained in the Show with Friesian and Boereperd classes being included. Dressage made a return despite the constraints of the grass surface with Kirsten Wing scoring a first for South Africa by riding the new M2 dressage test. The lower grade adult classes culminated with top-quality pony rider show jumping and good entries in all categories.

The now separate Royal World Cup Qualifier, which took place over five days in April 2009, attracted competitors from all over the country and the Main Arena was considered to be in 'excellent condition'. Dominey Alexander riding Alzu Barracuda won the qualifier class and Lisa Williams the Royal FEI International Victor Ludorum on Equi-Librium Warrangai P.

Pleasingly, that year the FEI awarded the World Cup Qualifier class four-star status but there was growing concern that in the prevailing economic climate it was becoming increasingly difficult to find sponsorship for what had become a major equestrian and social event. While the RAS was no longer willing to provide support in cash it was still ready to assist in kind with its facilities, free of charge.

In 2010 the Main Arena was again dressed with sifted sand and the warm-up area re-levelled under Rob Martin's supervision. The FEI World Cup Qualifier was reincorporated into the Royal Show after a two-year absence and Gareth Neill riding Mirage 85 won it.

The Association of the Friesian Horse Studbook of Southern Africa held its national championships during the Show with Jan Hendriks and Sjouke de Groot from Holland judging and Robyn Kenyon riding Reinoud fan de Homar, the winner. The Mounted Games Southern Hemisphere championships also took place then with two South African teams competing against those from Australia, Botswana, Zimbabwe and the winners New Zealand.

Sadly, the FEI competition was cancelled in 2011 after equestrian adjudicators decided that the Main Arena surface was too hard for A-grade jumping. The venue had been used for the UCI BMX world championships the previous July, which had necessitated the introduction of more than 12 000m^3 of sand to meet the required international specifications. Although this had subsequently been removed it was estimated that repairs to the unavoidable compaction which had taken place would cost an unbudgeted R105 000.

The initial assessment of the surface was subsequently overturned by other expert opinion, but too late to amend the altered programme to include those A-grade jumpers who were still willing to compete. Show jumping was by then becoming an increasingly expensive activity and the horses involved were each worth a fortune. This unfortunate episode pointed to the inescapable fact that the Main Arena was a multi-purpose facility whose surface could not be maintained exclusively to meet the high expectations of all A-grade jumpers as at dedicated venues like Shongweni near Durban.[21]

After Showtime and the RAS drifted apart, largely over funding, and with the assistance of Colin Scott, Jacquie Pappas and others, the equestrian programme at the Show began to take on a new lease of life. This was designed to change waning public perceptions of equestrian shows by including an entertaining and informative mixture of events.

The Society was still anxious to provide 'a meaningful home' for the 'equestrian fraternity' both in and out of Show periods. Following discussions with representatives of the Natal Horse Society the RAS donated R12 000 to it which was to be spent equally on showing and eventing disciplines.[22]

The Show began to emerge as a major, if not the country's premier, venue for the showing classes. In addition to the traditional showing of quality horses and the usual breeds, miniature Falabella and imposing Friesians were now also included. Carriage driving demonstrated how an old mode of transport had become a modern sport while reining showed how a horse and its rider can work as a team.

There were mounted games in which pony clubs competed in a gymkhana and cross-country jumping with teams comprising two adults and a child competing for a R9 000 prize. Double slalom was introduced involving 32 horses participating in pairs over six jumps and sixteen races in a knock-out format.

The country's leading horse racing operator Gold Circle also participated in collaboration with the South African Jockey Academy to produce a horse and harness racing display and a fixed exhibition designed to advise youngsters about career options in the sport. No prizes were listed in the RAS annual reports from 2011, but public interest was again gathering momentum even if not to the level of the 1960s and 1970s.

The **Crafts and Home Industries Section** was as busy as ever, displaying 3 500 items in 2002 submitted by 530 exhibitors, many of them youngsters. By then the section had long-since laid claim to being one of if not the

The Crafts and Home Industries Section enjoyed experienced and dedicated leadership with its former secretary **Di Fitzsimons** serving as its president (1998–2008). Since childhood the Royal Show had been a 'normal part' of her life as her family travelled every year from Zululand to Pietermaritzburg during Show time when they stayed with her uncle A.C. (Bertie) Bircher, RAS secretary (1940, 1945–1968) and then general manager (1969–1973).

In 1980 she and her husband moved from Empangeni to the city where Beth Upfold invited her to serve as a steward in the Crafts Hall. She subsequently joined its committee and then became an assistant to Mavis Fowle whom she succeeded as secretary. Di Fitzsimons continued her active involvement in the section for another decade following her subsequent term as president.[23]

county's premier crafts exhibition, including exhibitors representing the full demographic and age range of the region.

From 2002 the Crafts Hall boasted its own Station Café, so-called because the building shook every time a train rattled by on the Victoria Siding behind it! Thanks to hardworking volunteers working shifts, this provided an added attraction and another source of funding to supplement what Di Fitzsimons recalled was a very small budget and only limited sponsorships.[24]

In 2003 that section hosted the Federation of Women's Institutes, which celebrated its 75th anniversary. It also marked an unbroken partnership between the two during that whole period. Thereafter the various branches of the federation continued to contribute to the extensive range of activities and hobbies represented at the Show. These included tapestries and needlework, embroidered saris, painting and photography, woodwork, porcelain dolls, preserves and confectionery.

Occasionally honorary exhibits were hosted, such as a display of embroidered ecclesiastical robes and another of intricate headdresses that well-known milliner Lyn Douglas had designed. In 2006 the focus was on the World of Beading to which KwaZulu-Natal Museum Services and the Voortrekker Museum contributed ethnic and Victorian examples.

The following year the emphasis was on an exhibition of more than 100 quilts, some produced by development and outreach groups as the section sought to become more representative of the region's demographic profile. Entries continued to increase thanks largely to Di Fitzsimons' enthusiastic committee.

In 2008 the section celebrated its centenary with a programme of additional activities and no less than 3 650 exhibits, which was the highest in a decade. The *Witness* sponsored a photographic competition in memory of Clint Zasman, a former employee who was tragically killed. There was an archival display of old letters, minute books and photographs as well as honorary exhibits of the work of members and former members.

The various women's institute banners added colour to the occasion while a centenary luncheon brought together current members and participants, some of whom had been involved for more than 40 years. It was a 'fitting finale to a century of dedication to the Royal Show'.[25]

The whole section underwent a general revamp in 2009 to refresh its image. Under the *Daily News* Crafts in Action banner, it expanded into the adjacent Hall 4. Directed by Bobby and Lyn Hex as well as Carmen Nänni, it included a retail and interactive demonstration component at which visitors could witness various crafts at first hand.[26] Entries in all categories, particularly photography, increased in 2010 and a full display cabinet was devoted to the hand knitting of Lorna Oliver who had been a regular participant since the age of 10. Sophia Fiddes won the crochet category for the 15th year in succession, but sadly died a week before the Show.

In 2011 the Crafts and Home Industries Hall itself had to undergo a revamp, costing R40 000, to eliminate borer from its woodwork. After the experimentation of the previous two years the commercial craft displays were again separated from the competitive categories. Just Boutique Events and the Roseway Waldorf School complemented the commercial exhibits with demonstrations and workshops which included several other schools. These featured skills ranging from beading, icing and woodwork to doll making, pastels and sand art. There were seventeen exhibitors in the new Crafts in Action Expo which included foil art, fabric painting, scrapbooking and art produced from recycled goods.

Among the section's major prizewinners during this period were what Di Fitzsimons described as 'the remarkably talented' Ruth Aldridge (formerly Suttie),[27] Mrs Hockly and Una Taylor who won the senior championship trophy in 2004, 2005 and 2006–2007 respectively. Amy Grant was junior champion in 2004 while Thomas Rich, junior champion in 2006, won 'specials' and/or 'firsts' in four different divisions and was probably the first boy to win the title.

The Richmond and Jubilee Women's Institutes and Mooi River won the F.K. Lawson Shield in 2004 and 2005 respectively as did Richmond (again) and Twini Park in 2006, Mooi River (again) in 2007, Estcourt in 2008 and Swartkop Valley in 2009.

Newton High and Laddsworth Primary were senior and junior school champions in 2005 and again in 2006, 2007 and 2008. In that, the section's centenary year, Ruth Aldridge won the senior championship for the 13th time in her career and again in 2009. Emily Burger from Maritzburg Christian School won the junior championship with Wartburg-Kirchdorf and Newton High emerging as school champions. There was a noticeable decline in interest among the youth and women's institutes, although by then the section was widely regarded as probably the best exhibition of its kind not only in South Africa but in the southern hemisphere!

Regular participants Una Taylor and Ruth Aldridge were senior champion and runner-up respectively in 2010, Emily Burger again won the junior title with Cowies Hill winning the women's institute competition, Laddsworth the junior schools and Newton High again the senior schools section. They did so yet again in 2011, with Ruth Aldridge runner-up to Avril Musgrave as senior champion, Matthew Burger the junior champion and Northdene the women's institute winners.

On one occasion a new volunteer steward in the section briefly achieved similar success in the needlework category before it was established that the items she submitted for judging had been purchased from a shop to which the previously successful Ruth Aldridge (Suttie) had sold her uniquely distinctive work! The new section chair Henrietta Whelan effectively dealt with this awkward situation by asking the miscreant to identify (unsuccessfully) the seams she had used.

It was reminiscent of much earlier unfortunate incidents. On one occasion it was strongly suspected that Koo brand peaches had been entered as a home-bottled preserve. On another a stuffed black-velvet pussy cat was displayed after being sold to a shop in Durban by a previous entrant who was able to identify the materials with which she had made it.

Careful administration ensured that the overall incidence of dishonesty and theft was minimal. During the course of one Show a field mouse settled its offspring in a matineé jacket and at another some icing was eaten off a cake before the judging. The following year that entrant appropriately decorated her cake with miniature mice![28]

The **Commerce and Industries Section** was also highly competitive and continued to be unable to meet the demand for exhibit stands. It could therefore be quite selective in offering a balanced mix of goods and services available to the public.

The judging process was based primarily on that of royal shows in Australia, the Royal Queensland Show in particular, and was subjected to ongoing refinement. This ensured greater accuracy and fairness, but also provided a useful basis for selecting the exhibitor mix in subsequent years. By 2006 four external judges were adjudicating each stand independently on two separate days. The marks awarded within all criteria were then collated to provide the basis for a full report which was made available to exhibitors in the interests of transparency and possible future improvement on their behalf.

The growing number of foreign participants included the Belgian government which exhibited for the first time in 2003. Local blue-chip participants also increased, including Afrox, Barloworld, the National Prosecuting Authority and the South African Navy. An exhibitor survey in 2004 indicated that virtually all participants believed their involvement had yielded positive results. Among caterers the service clubs experienced a decline in turnover but, in common with shows abroad, brand name commercial operators enjoyed substantial profits.

By 2005 blue-chip participation had become the norm and included fourteen national and provincial government departments, the SANDF and KZN Wildlife as well as Caterpillar, Ford, Mercedes Benz, Mondi, the South African Post Office, Telkom, Toyota, UNISA and Vodacom. That year the SABC display comprised an impressive six-storey edifice including a stage, a retail marketing section and broadcasting facilities. These presented a full programme of entertainment involving several radio stations and SATV. It was one of the highlights of the Show and generated extensive publicity.

The following year ABI, Bell Equipment and Hulett Aluminium further strengthened the blue- chip component and the SAPS mounted an eye-catching display. There were no less than 415 commercial, industrial and service sector exhibits while several agricultural companies were denied participation for the first time due to the shortage of space and in the interests of maintaining a balanced mix.

In 2007 there was an increase in the representation of heavy construction equipment, including Babcock, New Holland and Longgong, while most tractor, implement manufacturers and car dealerships were still among the 420 exhibitors. They were all requested to comment on the extent to which their expectations of the Show had been met, with 20% reporting excellent, 60% good, 18% as expected and 2% poor.

The following year 59.6% responded that their expectations had been exceeded and 35.8% that they had been satisfactorily met. This was a pleasing outcome in view of the recessionary conditions the exhibition industry and the economy were experiencing. By 2008 the Commerce and Industries Section had a waiting list of 60 aspiring exhibitors while 22% of current participants were newcomers to maintain a good mix while still introducing changes.

There was further expansion in the exhibition of construction equipment, heavy vehicles and materials handling items with Doosan, JCB, Toyota forklift trucks and Volkswagen involved for the first time. Hall 1, dedicated to tourism, the Midlands Meander, the Albert Falls Amble and Sani Saunter, attracted considerable public interest. So did Hall 8 which was focused on security and agricultural exhibits with the SAPS display winning a gold medallion and the Royal Commercial Trophy as the best in any exhibition hall.

There was, however, growing concern about the 990m² Hall 2 which, under the auspices of the Pietermaritzburg Chamber of Business, fell short of expectations and lacked appeal judging by the low foot traffic through it. In

2009 it was converted into the Natal Witness Beautiful Homes Expo dedicated to home decoration and renovation.

Hall 1 focused on small, medium and micro-enterprises with Hall 8 still on safety and security: Business Against Crime, Correctional Services, the Director of Public Prosecutions and the SAPS all presented award-winning displays. The participation of construction companies declined as the economic downturn took effect, although some agricultural implement manufacturers like Massey Ferguson were requesting larger stands. Overall visitor numbers to the Commerce and Industries Section increased and continued to do so in 2010.

Further improvements to the impressive 21 000m² Food Hall during this decade enhanced its public appeal. These included the transformation of the adjacent Fraser Room into an impressive demonstration kitchen in collaboration with Capital Kitchens, Hirsch's and Inovar Flooring. In 2002 the South African Chefs Association, in partnership with Spar, was the first to demonstrate the potential of this facility. The following year, as a corporate sponsor and exhibit organiser, Spar assumed naming rights to the Food Hall.

It continued to gain in popularity with nearly 110 000 visitors passing through it in 2003, making it one of the biggest dedicated exhibits of any kind in the country. In 2007, with the generous sponsorship of Ilovo Sugar, the demonstration kitchen underwent a further complete revamp. While the quality of all the displays continued to improve, they were still not considered to be on a par with the 'exceptional standard of the Royal Highland Show' which was used as a role model.[29]

The whole section was grateful to its various sponsors, including FNB and the Pietermaritzburg Chamber of Business. In 2005 the *Mercury* was awarded naming and organisational rights to the Food Hall and adjacent demonstration kitchen, giving the Show additional media coverage. By 2009 the Mercury Hall of Food and Fine Living, together with what was now the Illovo demonstration kitchen, were as popular as ever and by 2011 attendance had continued to increase.

Rivalry for the trophies presented each year in this section was unsurprisingly strong considering the competition to acquire and keep a stand in the first place. Mondi won the Royal Commercial Floating Trophy for the best display in an exhibition hall on three occasions (2004–2006) and Telkom twice (2007 and 2009).

The University of KwaZulu-Natal (UKZN, formerly the University of Natal) won the Royal Industrial Floating Trophy for the best display in an individual building from 2002 to 2004 inclusive, after withdrawing its exhibit in 2001 but returning when offered a permanent and more prominent site at the main entrance.[30] Medi-Clinic (Pietermaritzburg) won that award from 2006 to 2009 inclusive but withdrew its exhibit in 2010.

Weston Agricultural College won the Hulett Aluminium Floating Trophy for the best display on an open site in 2003 and 2004 while the SANDF won it in 2006 and 2009 and Husqvarna South Africa in 2008 and 2010. Northmec won the Pietermaritzburg Chamber of Commerce Agriquip Floating Trophy for the best display of agricultural equipment in 2002 and 2011.

The Pietermaritzburg Sakekamer Shield for the best display by a new exhibitor was not awarded every year, unlike the President's Trophy for special endeavour which the SABC won in 2002, 2003, 2004 and 2006. Toyota SA Motors won it in 2005 and 2008 and the SANDF in 2010 and 2011.

There were several other annual awards in this section with Bell Equipment winning the Pietermaritzburg Chamber of Commerce and Industries Floating Trophy for the best South African industrial display in 2008 and 2010. The Gutter Company won the Chamber's Award for the most outstanding trade exhibit by a member of the private sector based in the greater city area from 2008 to 2010 inclusive, having previously won the Pietermaritzburg Chamber of Business Trophy awarded to a small business exhibitor in 2006 and 2007. Toyota SA won the Wesbank Trophy for the best motor industry display in 2006–2007 and 2010–2011.

Additional attractions at the Show continued to be interesting and varied. When in 2004 KZN Ezemvelo Wildlife gave up its external stand and donated it to the Society, UKZN's Faculty of Agriculture stepped in to set up a crowd-pulling animal nursery and birthing exhibit. The Asian Bazaar established along Shark's Alley on the south side of the Main Arena effectively revitalised the entire area.

For a time, the R106 838 the South African Music Rights Organisation (SAMRO) demanded of the RAS in 2006 threatened to silence the music played via the public address system during both the Royal and the Garden and Leisure shows.[31] In 2008 a longstanding feature of the Show that was popular among adults, Rotary's Foaming Tankard pub, closed. From 2009 the small modern funfair relocated to the east of the Main Arena to become another attraction for youngsters. It was extended in 2011 to include more fun rides by demolishing the ablutions that had previously served the pub.

As before, the Royal Show was not the only important function on the Society's annual calendar.

The **Garden and Leisure Show**, now firmly established as the country's leading horticultural event, was pre-eminent among the various other activities taking place on the Society's premises outside Show time. In September 2001 it was held for the first time in conjunction with Independent Newspaper's *Sunday Tribune* as its new naming rights sponsor. It subsequently confirmed its commitment through to 2006 inclusive and again in 2007. The greater publicity this association afforded soon exceeded expectations.

Attendance was always subject to the vagaries of the weather, but there was a noticeable increase in visitors from outside the province. A successful innovation in 2002 was the introduction of a Garden Show or Green Train that was fully booked and transported 144 enthusiasts from Gauteng to Hilton Station, which was spruced up, along with its steam train museum, to receive them.

There was also significant growth in exhibitor participation. The Bloemfontein Parks Department, whose 2001 display was remembered as 'stunning', was followed by its Pretoria counterpart, as well as a first-time corporate presence on the part of Blomeyers, Hirsch's, Nedbank, Standard Bank, Toyota and Vodacom. They were joined by others, including the journals *SA Gardening* and *Garden and Home.* In 2003 the latter described it as 'undoubtedly the top landscaping and horticultural show in the country'.[32]

That year, with FNB's generous support, a reconstruction of South Africa's exhibit at the Chelsea Flower Show was an added attraction. In 2004 the national Orchid Show and Congress drew participants from as far distant as Japan and the United States. The guest of honour, Dr Henry Oakeley, executive member of both the Royal Horticultural Society of England and the Chelsea Flower Show, observed that the 'stature of the Maritzburg event is on the cusp of achieving international status, with the top exhibits being certain medallion recipients at Chelsea'.[33]

Local judges contended that as many as ten qualified for that category. *Garden and Home* and Msunduzi Municipality shared the Best on Show accolade with both of them being awarded gold medals. So too, among others, were the Midlands Rose Society, eThekwini Parks Department, the *Sunday Tribune* and Mogale City (Krugersdorp), with the last being the best first-time exhibitor. The Show still clearly retained its status as the country's largest three-day show of any type and, as a further marketing tool, a four-minute video and CD based on the 2003 show was commissioned.

Despite dreadful weather in 2005 the overall standard of exhibits was considered to be on a par with the previous year. Cape Town landscaper David

Davidson, the current designer of South Africa's annual Chelsea exhibit, was the guest-of-honour and Pietermaritzburg's Pot-Pourri Garden Club of amateur enthusiasts outdid the professional competition to win the gold medallion with laurel.

The up-and-coming eThekwini Parks Department's display was adjudged the best keynote garden; and the Nelson Mandela Metropole's the best new garden. The province's grade 7 school classes were invited to compete in producing 4m² mini gardens for the first time while the floral cake icing competition, now deemed the best countrywide, enjoyed the first-time sponsorship of Ilovo Sugar.

In 2005 the RAS became an affiliate member of the Royal Horticultural Society, leading to closer productive interactions. By 2006 the Garden Show was being described as 'the continent's oldest and largest horticultural event', as 'South Africa's Chelsea' and as the most prominent event of its kind in the southern hemisphere, with increasing numbers of visitors from outside the province and from abroad. There was still a perceived need to market the event more effectively in greater Durban, Johannesburg and Pretoria.[34]

For the first time. visitors were invited to vote for the best People's Choice exhibits. They ranked the top five as eThekwini Parks Department, the *SA Gardening* Herb Garden, the Chelsea exhibit, *The Gardener* and Mbombela (Nelspruit) Municipality. The judges awarded three of these (excluding the Chelsea and Mbombela displays) as well as Northern Park Primary gold laureate medallions for being evaluated at 95% or above. They distributed 27 gold medals in all.

That year there were 192 commercially connected exhibits as well as four keynote gardens with displays varying in size from the school mini gardens to the 300m² allocated to local authorities and corporate entries. Mbombela, Tshwane (Pretoria) and Buffalo City (East London) were all new municipal exhibitors while the 2006 national Rose Convention, held in a 1 000m² Hall of Roses, was an important addition.

So too were the prestigious Institute of Environment and Recreation Management conference and the South African Flower Union congress held in 2007, with the former attracting 200 delegates from all over the subcontinent. The latter was accompanied by an impressive 700m² display while the re-creation of South Africa's latest award-winning national Chelsea Show exhibit from Kirstenbosch was another crowd-puller. There were 49 keynote gardens that year with the judges and the People's Choice both acclaiming eThekwini Parks as the best display on show.

It was decided to eliminate the words 'and Leisure' from the name of the Show to emphasise that it was now firmly established as a major horticultural event, even though there would still be room for leisure and lifestyle exhibits. Prunella Scarlett, a senior Chelsea Flower Show adjudicator, attested to its international standard in 2009 suggesting only that local designers needed to give more attention to foliage quality.

South Africa's 2008 Chelsea Show exhibit was carefully reconstructed as was fellow Chelsea medallion winner eThekwini Parks. Although poor weather affected attendance many of the 183 commercial exhibitors enjoyed increased sales while caterers and food vendors experienced a 15% increase in turnover. For the 30th occasion, the Show hosted the floral cake icing competition as well as a baroque concert, five 'ready steady plant' demonstrations and the provincial dog agility and jumping championships.

The 2009 Garden Show was favoured with ideal weather that helped to boost attendance. As usual the majority of visitors were drawn from the greater Durban area with about 2 800 from outside KwaZulu-Natal, judging by a survey of car registration numbers conducted in the parking areas. There were more than 350 display gardens, many emphasising self-sustainability with the inclusion of fruits, herbs and vegetables.

After several years of involvement, the national Chelsea exhibit did not feature in 2009 following FNB's withdrawal of sponsorship. Instead, the highlight was a reconstruction of the eThekwini Parks entry which won a gold medallion at that year's Chelsea Flower Show. It was the People's Choice but, as an honorary exhibit, the Chelsea-mentored judges, Arthur Mennigke, Jan Blok, and Terry van der Riet, did not consider it, instead rewarding the *Sunday Tribune* Garden.

The latter's designers, Sue Tarr and Mike Hammersley, achieved a remarkable 100% – a record in South African horticultural history. This from a panel of adjudicators over whom the RAS exercised no influence, having long-since divorced itself from the judging process in the interests of complete impartiality. Unfortunately, for all its success, the 2009 show incurred a R56 079 loss which pointed to the need for imminent changes.[35]

When the Garden Show celebrated its 35th anniversary in 2010 the event was widely considered the best to date. It attracted 21 000 visitors (10% more than the previous year) with an estimated 4 000 from other provinces and abroad. The first-day Friday attendance of 8 905 fell just short of the record 9 007 set in 1982 when the event was still a novelty.

Among the now numerous corporate and municipal participants were first-timers Cape Town Municipality and Synergy Scapes, adjudged the best newcomer. eThekwini, eligible again in both categories, was both the judges' and the People's Choice as the best on show although it was one of the small garden club displays, from Cramond, that scored the highest points of all the participants.

The centrepiece of the show, as always, was the *Sunday Tribune* Parks Hall. Its entrance featured appropriate displays from various museums and private collections to celebrate the 150th anniversary of the arrival of the first Indians in the country. Mr Anil Sharan, Consul-General of India, officially opened this exhibit.

The daily 'ready steady plant' contest continued to attract public interest, as did the invitation to the public for twelve contestants to participate for the first time in developing 4m² mini gardens on site. In addition to these and the corporate entries there were 200 displays of varying size, some involving participants younger than ten years old.

There was some discussion about not holding the 2011 Garden Show in view of the distraction of the municipal elections and the current 'parlous state' of the horticultural industry. However, the major stakeholders all showed willing and, in addition to the usual advance publicity in the print media, the show featured on radio and on Tanya Visser's DSTV programme 'The Gardener'. In all, it was estimated that it benefited to the value of R2 million by way of publicity.[36]

Despite unkind weather there were 258 gardens on display. These ranged from school mini trays to 61 entries which were 4m² or more in size and some bigger than 200m². In addition to the numerous novice and non-professional participants, including garden clubs and mini public gardens, there were several major feature gardens from different parts of the country. Cape Town's exhibit was the People's Choice and the *Sunday Tribune* Garden by McDonald Stuart Landscaping won the official judges' award.

Anja Taschner of Ludwig's Roses oversaw the launching of a new rose *Rosafrica*. The show also featured the country's leading floral icing display competition which overall won gold with laurels for the first time. The guest of honour, prominent horticulturist Keith Kirsten, gave a series of talks, there were culinary demonstrations, dog agility trials, a 4x4 course and a display of caravans, boats, trailers and vehicles.

Other activities which, in addition to the Garden Show, made use of the Showgrounds were growing in number and collectively contributed increasingly to its financial wellbeing. In 2002 several blue-chip events were held including the launch of new Chrysler, Daimler and Volvo models.

The national Department of Transport chose the Olympia Hall with its much-improved ventilation for its 3 000-seater conference venue. The same hall was used for the 2 300 exhibits presented at the highly successful July 2004 Southern African Poultry Championship Show. Other poultry, budgerigar and rabbit exhibitions followed regularly in subsequent years.

After an intensive 2004 marketing campaign the use of the Society's facilities outside Show periods increased by 30% and as much as 150% in revenue terms. This included the three-day opening of the 2005 session of the provincial parliament in the Olympia Hall which subsequently became a regular event. Outside, Toyota SA launched its new range of bakkies and bussed farmers in from far and wide, following this with a display of off-road vehicles in 2007.

There were nine equestrian events in 2005 followed by many more in subsequent years, all of which culminated in the Royal Show. This suggested that the Showgrounds were still the preferred venue for such occasions despite the availability of several alternatives. From 2006 the Main Arena, jump equipment and stabling was made available free of charge to them. The following year the RAS hosted the junior and adult Midlands championships as well as the inter-schools show jumping competition and the highly popular Pony Show.

In 2006 the revenue generated by these and other activities, including an Elvis Tribute, Science Tech Fair and Cheese and Wine Fair, more than doubled. By then more than 12 000 visitors a month were attending exhibitions, conferences, formal dinners, weddings and other private functions, UKZN graduation ceremonies, product launches, business seminars, and poultry and rabbit club shows.

This generated at least R748 000 in services alone while the Society's premises had become second only to Durban's ICC as the busiest and largest general function venue in the province. The shortage of sufficient short-stay accommodation in Pietermaritzburg had suddenly become an impediment to further growth. RAS president Andrew Line attributed this increase in the use of the Showgrounds and the additional income it generated largely to Terry Strachan's strenuous marketing efforts in what seemed to be an era of conferencing.

Among other events taking place outside Show time were the Natal Youth Shows hosted in January 2006, 2007 and 2008. These enabled it to re-establish itself as a popular educational occasion and become more representative of

A.J. (Andrew) Line served as the Society's president (2004–2006) in the footsteps of his great-grandfather Leonard Line (1915–1929) and father John Line (1973–1981). His grandfather Richard George Line had served on both the general and executive committees, but was a retiring individual who declined the presidency. In his younger days Andrew Line became very familiar with the Showgrounds through regular visits to the Royal Show because his father owned a farm in the Dargle, exhibited Ayshires and became a judge of them and of Frieslands.

In the late 1990s the president Derek Spencer encouraged Andrew Line to join the RAS committee structure on which he served for six years before becoming president, on the understanding that he only wished to serve one term. In common with other occupants of that office Line remembered it as an interesting if sometimes stressful experience and involved interacting with a variety of personalities. He added a valuable business dimension to the executive which was always welcomed to balance the ever-important input of those directly engaged in agriculture.[37]

the province's demographic profile, thanks in part to the initial involvement of Zhake Agricultural College. Judy Stuart was largely responsible for ensuring that exhibitors from Qwa Qwa participated in the Shows but it was recognised that more young people needed to become involved.

In 2009 as a trial run the UCI BMX supercross world cup was held in the cattle arena followed in 2010 by the BMX world championships in the Main Arena. The latter event attracted more than 25 000 attendees, many from overseas, as well as wide media attention including more than 1 000 hours of international TV coverage.

The RAS sought assurances from the organisers concerning possible damage to its property. As previously mentioned, it paid the price for having its Main Arena surface declared no longer suitable for A-grade show jumping. From time to time there were other unbudgeted expenses incurred such as the damage caused to the cattle arena by the extended stay of animals on site following the nGuni Club sale in April 2011.

Nevertheless, despite the prevailing recessionary climate, by 2011 the overall use of the Showgrounds' facilities other than in Show periods contributed no less than 22% of its revenue. As early as 2006 it was recognised that the Royal Show, the Garden Show and other use of the premises had indeed become 'the three branches of the Society's business' when, for the first time, the third of these contributed more than R1 million to its income.[38]

Moreover, with a collective attendance of 180 000 at such functions in 2010–2011 the Showgrounds could now claim to be the most heavily used conference and function venue in KwaZulu-Natal. Its location, reasonable rates and provision of everything that was required for such events gave reason for optimism that it would continue to be much in demand. The same could not be said of Society membership.

Membership of the RAS, as distinct from attendance at the annual Royal Show and the Garden Show, was still drawn from the white ethnic group despite efforts to broaden its demographic base. Partly for that reason, considering that whites constituted a diminishing percentage of the overall population, membership continued to decline gradually from 2 927 in 2001 (2 444 adults, 398 juniors and 85 life members) to 2 474 in 2006 (2 048, 345 and 81 respectively) and 1 535 in 2011 (1 296, 158 and 81).

Other factors probably contributed to this trend, including recessionary conditions that reduced the discretionary funds of many households. Efforts to attract new members through local newspaper advertisements, from educational institutions such as Cedara, Weston and the local university, as well as by waiving the annual membership fees of paid-up members who introduced a new adult member, did have a temporary positive effect.

Apart from the minimal income derived from membership fees there was concern that the ongoing decline in numbers could eventually 'complicate the Society's constitutional raison d'être'. It was suggested that membership should become obligatory for entrants in the livestock sections as was the case at the Pretoria Show.[39]

Attendance at the annual Royal and Garden shows, by contrast, was affected as always by prevailing weather conditions and other factors that varied from year to year.

The weather was particularly unfavourable during both events in 2002 when attendance at the Royal Show dropped by 20 902 whereas the unusually high attendance of 225 660 in 2005 included 29 000 school pupils drawn from all over KwaZulu-Natal on a single day. This was almost certainly a national record and served to demonstrate the Show's broad though fluctuating demographic appeal. Requests from some exhibitors to limit the attendance of schoolchildren were declined on the grounds that their admission could not be reduced any further and that school groups were currently confined to Wednesdays and Thursdays.[40]

The 15 000 decline in 2008 to 155 475 was due almost entirely to a drop in school pupil numbers and in attendance at the Ukhozi concert on the final Sunday of the Show. Both were attributed to rising fuel costs. When, in 2009, the concert was held a week after the Royal Show, devotees declined from 27 241 to 12 214 although overall attendance at the Show increased slightly. It improved again to 136 031 in 2010 despite another decline in pupil numbers, but this included the 21 000 who were attracted to the Ukhozi concert that had been reincorporated into the Royal Show. Overall, attendance at the Show declined from 228 813 in 2001 to 161 057 in 2006 and 146 312 in 2011.

Compared to the more varied ten-day Royal Show, the shorter three-day Garden Show with its perishable exhibits was even more at the mercy of the weather. Consequently, attendance fluctuated unpredictably from year to year, rising from 21 114 in 2001 to 23 050 in 2006 but declining spectacularly to 16 734 in 2011. Deteriorating economic conditions doubtless contributed to these adverse trends. This also had an impact on the Society's finances.

Finances were indeed heavily dependent on gate and parking income, which still amounted to 33% of revenue in 2011 compared to 43% in 2001. Show stand rentals generated 28% of income, compared to 31% a decade earlier, and general use of the grounds accounted for 22% compared to a mere 7% in 2001. Only 1% came from membership fees in 2011 compared to 2% in 2001.

As before, the major annual expenditure was on salaries and wages, holding steady at 25% of costs in 2001 and 26% in 2011, followed by the general category of other at 19% declining to 18% a decade later. There were few significant changes in the other usual expenses, with advertising rising from 9% to 11% and municipal costs from 6% to 8% while repairs and maintenance were reduced from 10% to 7%.

Following the recent upward trend there was a moderate correction to the Society's financial situation in 2002 due partly to changes in the way it recorded depreciation. With the exclusion of that factor the annual balance sheet actually amounted to the second best in RAS history while 2003 proved to be its most successful ever. There was a 53.9% increase in operating profit and a 35.2% increase in net surplus prior to factoring in depreciation and financial transfers to various reserve funds. This was followed by figures of 25.4% and 9.5% respectively the next year and another set of 'best financial results ever' in 2005.[41]

The executive tried, unsuccessfully, to negotiate a partial waiving of the rates levied on the land which the Society leased from the municipality. This was particularly frustrating because these had been imposed without any prior warning and contrary to the understanding reached in that regard in 1997. In 2002 a Matrix survey established that its various activities created some 2 000 jobs in the Pietermaritzburg region. Nevertheless, the following year, for the first time, the RAS had to pay 100% of rates due, amounting to R130 000.

However, not for the first time, the provincial Department of Agriculture acknowledged the Society's 'most acceptable' efforts to promote agriculture with a R10 000 grant. This was followed by R30 000 in 2003 and R40 000 in 2004, after which these grants came to an end. By then the rates account had risen to R474 000, including an additional R17 399 following a R658 000 estimated increase in the value of its buildings. Good news that year was that SARS confirmed the Society's tax-exempt status. This was subsequently carefully monitored, but was not to last indefinitely.[42]

In 2003 members of Manco were involved in discussions that envisaged an 'all-encompassing development in the Ashburton area adjacent to the N3 motorway' that would include making land available to the RAS. The proposal did not materialise and rumours that the Society was relocating to that site had to be disabused.

Instead, in 2004, following two years of negotiations which Terry Strachan conducted, assisted by the president Andrew Line and accountant David Wing, an agreement was reached with the Msunduzi Municipality whereby the RAS would relinquish its 50-year leasehold rights to the Chatterton Road stables and the neighbouring driving range (4 hectares).

In return the Royal Mews Trust would pay R6 million, which the Society would guarantee, in exchange for freehold title to the rest of the leased portion of the Showgrounds amounting to 13.5 hectares. It was subsequently decided that the RAS would raise the necessary loan and increased this to R8 million to cover transfer duties and legal fees. All the land in question was initially transferred to it but for tax purposes the rezoned node between Hyslop and Chatterton Roads was subsequently transferred to the Trust while the rest of the property remained under the Society's ownership.[43]

This arrangement promised much greater security of tenure than ever before in its 105-year history but it also implied further space constraints, not least with regard to stabling and the continued survival of the funfair. The latter's demise seemed likely to affect the popular appeal of the Show and reduce attendance. Nevertheless, the Chatterton Road stables and the existing funfair both ended with the 2005 Royal Show. After lengthy delays the transfer of freehold rights to all previously leased parts of the Showgrounds eventually took place on 24 August 2006.

It was certainly one of the most important milestones in the Society's history and one of the more significant achievements during Terry Strachan's tenure as CEO. The executive now faced the task of servicing the substantial loan, which FNB provided at a favourable 1.5% below prime, and embarking on a carefully financed schedule of refurbishments while rebuilding its reserves. A redevelopment of the funfair site was clearly a priority to offset the R6 million loan.

The municipality did reimburse the RAS R120 000 in lieu of salvaging re-usable materials from the demolished stables. Sixty new stables were built in 2006 within the Showgrounds at a cost of R600 000 to ensure that equestrian events in and out of the Show period were not unduly disrupted and that an equestrian presence (and membership) was maintained.

As anticipated, overall attendance at the Show dropped from 225 660 in 2005 to 161 057 in 2006 due in part to the closure of the funfair. Moreover, the delay in transfer meant that the attendant loss of R1.2 million in revenue could not be offset by renting that site to a new developer although several options were under consideration. This had adverse consequences for the Society's overall financial returns for 2006 even though its three primary sources of income – the Royal Show, the Garden Show and other uses of the grounds – were more than satisfactory.

By 2007 the RAS had clearly weathered the disruptions of the previous year and recorded a 101% improvement in its bottom-line figures, prior to allocations to special reserves, despite interest payments and rising costs. It had already reduced its FNB bond by R1.5 million, followed by another R1.1 million in 2008. A new, smaller funfair was opened within the Showgrounds which was well received by the public.

In 2004, in an effort to expand its territory further, the Society acquired all but one of the properties adjacent to the Showgrounds south of Hyslop Road for R1 750 000. This provided sufficient space for the replacement of stables and cattle stalls lost following subsequent redevelopments elsewhere.

Thereafter the RAS secured first refusal rights to the only remaining property south of Hyslop Road which it did not own. An agreement of sale involving R910 000 was concluded in 2010 following the death of the owner. This gave the Society freehold ownership of the whole block bordered by Howick, Hyslop and Chatterton roads.[44]

In addition, an offer was submitted to Spoornet to buy the Victoria Siding to the south of the Showgrounds as well as the triangle of land behind the Crafts and Home Industries Hall. In 2010 Spoornet eventually agreed in principle to sell the latter for R575 000 but not the former. This was never consummated.

The decline in gate takings at the Royal and Garden shows in 2008 was due to the drop in pupil attendance in the case of the former and inclement weather plus the distraction of the Rugby World Cup competition in respect of the latter. This was more than compensated for by income from the use of the grounds during the rest of the year, which improved by 34%. It helped to produce an operating surplus of R1.4 million despite the worldwide recessionary conditions.

This source of income increased further in 2009 when the RAS enjoyed a moderately successful year despite the decline in visitor numbers and the difficult conditions the international exhibition industry was experiencing. It necessitated renting out some Royal Show stands at discount prices and making grants to special blue-chip exhibitors to ensure their presence.[45]

In 2010 the Royal Show generated R600 000 more than in 2009 and after a lengthy dispute the municipality eventually agreed to reimburse a previous rates overcharge of R64 000.This income was, however, not sufficient to compensate for the significant decline in revenue the Society experienced from conference and function bookings as the economic recession took effect. Prudent control was necessary to achieve what was at least a reasonable set of year-end financial results in the prevailing circumstances.

The same was true in 2011 but although relations with senior civic office bearers had improved, the RAS suffered a setback when the municipality informed it that, in addition to a substantial increase in electricity charges, the special rebate it had enjoyed for 70 years was to be withdrawn. Following negotiations, the blow was softened when it was agreed to phase the grant out over three years.

In addition, for the first time the Society was also required to pay for the services of municipal pointsmen on duty prior to and during major events at the Showgrounds. Not least, recent amendments to the regulations suggested it might soon be regarded as liable for taxation. On 30 June 2011 the accounts of the RAS reflected a retained surplus of R25 850 558. Despite this apparently healthy financial situation the quest for sponsors remained as important as ever.

Sponsorships were difficult to find as the economic climate deteriorated during the first decade of the new century. This was despite the substantial captive audiences which the Royal and the Garden shows provided. Donations and sponsorships accounted for only 7% of the Society's income in 2011, dropping from 8% in 2001.

It was grateful for the support it received, in particular from its longstanding Royal Show corporate sponsors Vodacom and Coca-Cola (ABI) and from the SABC and Spar, as well as from the *Sunday Tribune*, Shepstone and Wylie Tomlinsons and Umgeni Water; and subsequently Huletts Sugar and Tomlinson Mnguni James in connection with the Garden Show. In 2003, Vodacom was granted naming rights to what had previously been the Stannic Tower which was then refurbished.

In 2004, Telkom joined the ranks of the Royal Show's corporate sponsors and provided the guest- of-honour at that year's opening ceremony. In addition, there were the aforementioned grants which the provincial Department of Agriculture made in 2002, 2003 and 2004 in recognition of the Society's contribution to agriculture.

Sadly, the liquidation of Stock Owners not only ended 'a longstanding friendship and association' for the RAS but also an important source of sponsorship, links with the farming community and logistical assistance, including professional stewarding, which would be 'sorely missed'. Its Showgrounds premises and sale ring left the Society with additional facilities for other purposes, but Stock Owners demise was a serious loss and deeply regretted.[46]

When in 2004 Umgeni Water withdrew its longstanding sponsorship of the municipal displays at the Garden Show the RAS had to meet the shortfall. In 2006, Standard Bank became one of the Royal Show's major sponsors and Illovo Sugar became a primary corporate supporter of the Garden Show, as did FNB the following year. In 2007, Standard Bank sponsored the first lavish Gold Cup gala dinner which replaced the earlier Wednesday afternoon Royal Show openings and trophy presentations that seemed to have lost their appeal and had attracted declining attendance. FNB continued to host the opening week breakfast.[47]

That year, after lengthy participation, Hirsch's and Mondi opted not to be involved in the Royal Show and the following year Umgeni Water withdrew its longstanding support for the riverside walkway whose maintenance became the Society's entire responsibility. However, Standard Bank, ABI and Telkom confirmed their sponsorship while Pioneer Foods also became corporate sponsors of the Show as did Liqui Fruit in 2008. Independent Newspapers (the *Sunday Tribune* and *Mercury*) remained major sponsors of both the Royal and Garden shows.

By 2010 it was reckoned that many sponsors and prospective sponsors had reduced their marketing budgets by as much as 70% in response to the unfavourable economic climate, but both the Royal Show and the Garden Show retained their major sponsors – ABI, Standard Bank and Telkom in the case of the former and the *Sunday Tribune*, Tomlinson Mnguni James and Illovo Sugar of the latter.

In 2011, Standard Bank withdrew but the Land Bank and provincial Department of Agriculture became major Royal Show sponsors. By then the corporate sponsors' dinner on the Monday evening of the Show had become a regular annual event by way of appreciation.

As before, the Society's financial well-being was also closely associated with that of the Royal Mews Trust.[48]

The **Royal Mews Trust** broke even financially for the first time in 2002 with both of its tenants, McDonald's Garden Centre and O'Hagans restaurant, seemingly now well-established. Unfortunately, the following year the latter's lease had to be terminated but a satisfactory new tenant was found in another restaurant trading under the name of Taps and owned by Ovenstone Industries with which a five-year lease agreement was negotiated. A sub-lease with the Firkin chain was subsequently approved as was McDonald's management contract with Dunrobin Nurseries.

Proposals to develop the former funfair site at the north-eastern corner of the Society's premises remained under consideration as discussions took place with Sedgley Developments and Barlow World. The latter subsequently relocated to the northern end of the polocrosse fields. A rezoning application to broaden that area was approved in 2003 which greatly improved the options open to the RAS with regard to its future. Ben Schutte, managing director of the funfair, welcomed the possibility of transferring it to a new site on the other side of Chatterton Road, but this was not to be.

In 2005 the trustees of the Royal Mews Trust and the Society's executive committee agreed that the 3.8 hectares of rezoned land on the Showgrounds side of Chatterton Road, extending from the former funfair to McDonalds Garden Centre, should be transferred to the Trust for R5.4 million. This was 'in the interests of preserving the integrity and separate entities of the Society and the Trust'.

Further, it was agreed that some income generating development should be considered for the funfair site with the Stock Owners Co-operative (prior to its liquidation) and Natal Agricultural Union being envisaged as appropriate key tenants. Late in 2005 an agreement was eventually reached between the Trust and Nick Proome representing Prospect SA Investment for the development of the site. Before year's end ABSA Bank had bought into the project and a second anchor tenant was being sought.[49]

Further developments had to be postponed due to the delay in transferring freehold title to the leasehold land from the city to the RAS, with adverse financial consequences for the latter. Also, transfer of the four separate sub-divisions could only be completed after all of them had been provided with independent reticulation services.

In 2007 the Royal Mews Trust enjoyed a net profit before tax of R427 627, the best results since its inception in 1999. With the sub-lessees Dunrobin Nurseries and Firkin & Ferret restaurant well-established, a 25-year lease agreement was concluded with property developers Prospect SA for the construction of a bank park on the funfair site. That company paid for the construction of the new 60 demountable stables in the old Block C Pig Section as well as a workshop and storage facilities.

After ABSA Bank contracted to assume part of the funfair site only two parts of the rezoned area remained undeveloped. These were the Sheep Section and the land occupied by three stable blocks and the workshop, immediately west of the ABSA development, each approximately 0.8 hectares in extent. It was subsequently decided that the Sheep Section should remain within the Showgrounds in the interests of the Society's primary agricultural function.

The Trust's R492 000 operating profit in 2008 was reduced to a R183 000 loss by the necessary interest payments on its R5.4 million land purchase and by funds held in reserve to cover a disputed late-year rates assessment. In 2009 the Trust produced a R304 978 surplus, its sub-lessees remained in good standing despite the difficult prevailing trading conditions, construction of new Standard Bank premises close to those of ABSA got underway and there were other keen developers asking for space there.

In 2010 the Trust enjoyed a further surplus before tax of R676 676 despite the current economic recession. It was sufficiently upbeat about the future to commission Matrix Project Consultants to assess the viability of establishing a hotel and conference centre in the area inside Gate 3A of the Showgrounds. This was considered to be viable subject to a partnership agreement with the city.

There was another before-tax surplus of R731 942 in 2011 although it was necessary to negotiate a new five-year lease with the Dros restaurant chain to replace the departing Firkin & Ferret and it was reported that the Dunrobin Nursery was now for sale. Both these developments were probably a reflection of the difficult prevailing business climate and a loss of R70 000 was incurred in connection with the former.

The ongoing survival of the RAS depended not only upon sound financial management, but also on the efficiency and loyalty of its employees.

Administrative and grounds staff employed by the Society numbered seventeen by 2001. They continued to play an essential role in ensuring the success of the annual Show and all the other events taking place in the Showgrounds. Staff changes included the departure of Jenny Perry (formerly van Niekerk née Fraser) who emigrated to Britain after nineteen years of loyal service, as did Carmen Paul after eighteen months in office.

The former observed on her departure that 'I do not know of any other organisation that has the ability to reach your inner being, squeeze your life blood, saturate your brain and then leave you feeling immensely proud to belong.'[50]

Margaret Mitchell joined the staff in 2002 as the CEO's personal assistant for what was also to prove a long term in office. That year Dix Reddy was formally thanked for his 'dedication and loyalty' to the RAS during the previous 25 years and in 2003 Kishore Singh left after seventeen years in harness to be replaced as grounds superintendent by Roy Motilal.

Murray Witherspoon (exhibition co-ordinator) and Barbara Shaw (accountant) also joined the team. Wendy Burnard left and with Tanya Leutenegger's arrival renewed efforts were made to market the use of Showgrounds facilities outside Show periods. She departed in 2004 and was replaced by Dee Newton. Ken Easthorpe's services as the Society's reliable longstanding electrical contractor were formally acknowledged at the 2005 corporate sponsors' dinner.

Relations between the staff and management remained very cordial and in 2004–2005 a wage agreement was amicably concluded, followed by regular biannual negotiations. Receptionist Rachel Bleakley retired after eleven years of service and was replaced by Penny Grobler. Simpson Bhengu was appointed following the death of tractor driver Mbukeni Sithole after twenty years of employment and Lucas Khumalo joined as a gardener. After nineteen years of service the versatile Leonard Mbanjwa suddenly resigned but his return was welcomed in 2007.

The following year Murray Witherspoon left for Johannesburg, Carmen Nänni (née Paul) returned to serve as exhibition co-ordinator and Roy Motilal successfully recovered from heart surgery. Marc Brooker joined the permanent staff but left shortly afterwards while Murray Witherspoon returned as head of front office and use of grounds.

Sheilagh Harman assumed the post of front office receptionist in 2008 and one of the supervisors, S.B. Bhengu, left. It was reckoned that on average the grounds staff had been employed for more than fifteen and some for more than 25 years. Dix Reddy, assistant grounds superintendent, celebrated his 30th anniversary with a long-service award.

Following Di Fitzsimons' suggestion that younger people should be encouraged to serve on the Society's Exco it was established that the average age of its fifteen members was 58, with only six of them being 50 or younger. In 2009, Manco accepted Kay Makan's more immediately implementable proposal that Terry Strachan's office as general manager should be changed to that of chief executive officer (CEO) in keeping with other member societies of the RASC and the responsible position which he held.

In 2010, in the interests of improving communications, the chairman of the Garden Show was co-opted on to the executive. So too was Marlize du Preez, having for 'several years been the sole driving force and inspiration behind the structure and growth of the Sheep Expo', as well as overseeing the attendance of school pupils at the Garden Show. It was anticipated that, at 33 years of age, she would add a much-needed 'youthful and dynamic element to the Society's governing body'.

Margaret Mitchell retired as personal assistant to the CEO in 2010 and was replaced by Lynn Falconer. The following year she and Murray Witherspoon left the Society's employ and Jenny Fraser (formerly Perry, formerly van Niekerk) rejoined the administrative staff.

Marlize du Preez completed a degree in Communication Science at the University of the Free State with an interest in event co-ordination and management. She was first invited to organise the RAS members' dinner and subsequently the combined members' and corporate sponsors' function. She was then asked to become involved in organising an expo for the Sheep Section in 2004.

This became her primary focus as far as work for the Society was concerned. The Sheep Expo was intended as an educational interactive display to illustrate the whole process from wool production to finished items under one roof. A challenge, among others, involved securing enough sheep for the ever-popular shearing demonstrations given the limited supply of animals in the KwaZulu-Natal region.

Marlize du Preez also became involved in the Garden Show by organising the school competitions producing mini-tray and later larger 'ready-steady-plant' gardens. These annually attracted between twelve and eighteen groups from as far afield as Ixopo and Hibberdene.[51]

In September 2011 the RAS recruited two additions to its grounds staff from the Howick SPCA in the form of mother and son donkeys Dolly and Dude. Their onerous duty was to roam freely around the property outside of Show time to assist in keeping the grass under control. They did not qualify for full benefits but were regularly (over)fed by the rest of the staff as well as becoming a familiar sight to visitors. They were accompanied by a turkey which joined the nine geese and six chickens in the Society's mini animal farm. By the end of the year this complement had increased to two turkeys, twenty geese and twenty chickens![52]

The **profile of the RAS** beyond the confines of the Showgrounds was projected at a rather more senior staff level at various relevant conferences and seminars both nationally and abroad. In 2002 Terry Strachan attended the biennial conference of the RASC in Northern Ireland which he found very valuable in sharing common challenges and potential solutions.[53] Andrew Line was accompanied by Dr Albert Modi of UKZN in representing the RAS two years later at the subsequent conference in Australia while honorary life president Derek Spencer did so in 2006 at Calgary.

The Society's profile was maintained in several other ways which attracted media attention. In 2004 bursaries were awarded to UKZN agricultural student Lyndon Gray and Cedara student Siyabonga Vilakazi, both of whom assisted at the Show. There were other recipients in subsequent years who were also given free RAS membership for the duration of their bursaries, two complimentary tickets to both the Royal and Garden shows and were invited to at least one major function.

In 2005 Derek Spencer was appointed a trustee of the RASC. That year the Society's nominee for the female Farmer of the Year award, Judy Stuart, who was a member of its cattle committee, secured second place. Garth Ellis attended the 2005 Kwanalu Congress and Iona Stewart the event held the following year.[54]

In 2007 Mike Moncur represented the RAS at the SAMIC awards dinner in Pretoria, in 2008 Iona Stewart, Garth Ellis, Mike Moncur and Terry Strachan did so at the presentations in Kempton Park and Stewart, Strachan, Marlize du Preez and Jonathan Tyler did so the following year in Vryburg.

In 2008 the Society co-sponsored Professor Sheryl Hendriks of UKZN as a keynote speaker on food security in Africa at the RASC conference in New Zealand. The RAS was also represented by Mike Frickel, Iona Stewart, Derek Spencer and Terry Strachan who gave a presentation on the role of livestock at future shows. The last three represented the Society at the conference held two years later in Edinburgh, together with Mark Stewart who was awarded a £500 RASC scholarship to do so as a next- generation delegate.

On his return he began to form a young RAS farmers' group dedicated to serving the Society in exchange for encouraging their development. Manco agreed that young farmers should be more actively involved in future shows and in 2011 a roster was drawn up involving several of them who volunteered to assist that year in the Cattle Section. A twenty-strong working group was subsequently given a talk on agricultural implements.

In 2009 the Pietermaritzburg Chamber of Business invited Terry Strachan to represent the Society at the Midi Strategic City Summit. It was highly appropriate considering the role the RAS continued to play in the city and region in creating jobs, attracting visitors and generating income. In 2011 Iona Stewart represented the Society at the Botswana Agricultural Show in Gaborone and at the Kwanalu conference.[55]

Iona Stewart was the first female president of the RAS (2010–2012) in its 159-year history. The first women had been admitted as members in 1909. She grew up in Johannesburg, graduated in 1962 from the University of Natal with a BSc (Agric.) majoring in animal husbandry and joined the Department of Agriculture in Pretoria. She subsequently relocated with her husband Greg, a fellow Natal graduate, to a succession of game parks in Zululand where he worked as an ecologist for the provincial Parks Board. There she engaged in research on impala before moving in 1966 to the USA where she and her husband completed Masters degrees.

On their return in 1968 she ran a dairy farm south of Pietermaritzburg for seventeen years before moving to a smaller property adjacent to the Cedara Research Station. In addition to farming, she worked at Cedara as a beef research specialist and completed a doctorate in animal science in 1996. The following year she was elected the first female chair of the Lions River Division Agricultural Society. After retiring from Cedara in 2004 she was able to concentrate on her nGuni cattle herd and devote more time to the RAS.

Iona Stewart's long association with the Society began in 1971 as a voluntary steward in the Cattle Section. In that capacity she learnt more about her field of expertise before becoming the section's chair and subsequently president. She also learnt that at Show time livestock exhibitors could become 'quite nervous', in some cases 'quite unreasonable, and actually very rude' although this was often followed by profuse apologies and even flowers![56]

Stalwart supporters and volunteers, like the permanent staff, were as essential as ever to the effectiveness of the Society's various activities and in maintaining its position as one the largest service organisations in South Africa. These included judges, marshals, stewards and numerous other helpers including the youngsters from Zakhe and Weston agricultural colleges who assisted with the livestock.

However, by 2002 there was concern about the diminishing numbers of volunteers in some sections of the Show, totaling only about 300 by 2009. The policy of granting annual honoraria to some but not all volunteers may have contributed to this trend by causing resentment and making it more difficult to attract sectional committee members without offering stipends.[57]

There was, as before, another way of acknowledging service to the RAS. In 2002 honorary life membership was bestowed upon Jenny Perry (formerly Van Niekerk née Fraser), the Society's departing administrative manager and secretary. She was followed by Sheila Dale, Ann Oldacre and Ronel Hojem, all of whom had contributed significantly to the Crafts and Home Industries Section.

In 2003 vice-president Frans Krause relocated to Pretoria after seven years on the management committee. He and Mrs D.R. Evans, a longstanding supporter from Richmond, were also accorded honorary life membership as were Mr and Mrs J.R. Torr. So too were Mrs Tim Hancock, Moyra Poole, Gordon Stewart and Una Watson for long service in various capacities, in the latter two cases for their contributions to the Garden and Leisure Show.

In 2004 Reeva Wing also became an honorary life member and her husband David an honorary life president while Rod Dorning, vice-chairman of the Sheep Section, died suddenly. The following year John Fowler withdrew from the executive committee after lengthy service and was formally thanked for his invaluable contribution to the Society. Ron Glaister replaced him as a trustee.

In 2006 Ben Schutte, who had run the funfair for many years, became an honorary member as did Ron Wright whose prize-winning photographs of previous Garden Shows came into the Society's possession. That year Royce Fann died after exhibiting and later judging canaries at the Show since 1948.

In 2007 outgoing president Andrew Line was elected an honorary life president while Wendy Line and Jenny Dalton were added to the list of honorary life members for longstanding services to the Society, as was Shaam Rajpul for his contribution to the carcass competition over several years.

They were followed by Pieter Breytenbach, David Hughes (posthumously, an executive member and for fifteen years the Society's architect), his wife

Pat Hughes and ex-councillor Osman Gani who had assisted in negotiating the transfer of freehold title in 2006. Other life member recipients were successful sheep exhibitor Russell Shorten, Dallas Kemp and Martin Seyffedt, who had contributed so much to the Cattle Section as commentator and judge respectively for more than 30 years, and Di Fitzsimons for her enormous involvement in the Crafts and Home Industries Section.

John Saville, who had been instrumental in developing the carcass competition, died as did Mark Shute, RAS general manager from 1974 to 1994. In 2009 Graham Atkinson was awarded honorary life membership for contributing to the development of the Society's Chatterton Road retail projects over a period of 14 years.

So too was Marie Andrews for 24 years (thus far) of unbroken service to the crafts and home industries committee. The Society's land surveyor Tom Trench and town planner Jan van der Vegte were similarly recognised for the advice they had provided over many years.

In 2010 Ron Wright died. Soon afterwards the Show tragically lost Graeme Pope-Ellis of Reapers Poldenvale who had regularly contributed valuable logistical support to the livestock sections. Barely three weeks earlier he had been accorded honorary life membership, as was Winston Moffett of Weston Agricultural College for overseeing the provision of student stewarding assistance in the livestock sections.

Iona Stewart and Mike Frickel were similarly honoured for serving on the executive committee for ten or more years and Graham Atkinson was further rewarded when he became an honorary life vice-president. Former president Garth Ellis was accorded the status of honorary life president and his wife Felicity became an honorary life member. In 2011 Margie Oliver, longstanding stalwart of the Crafts and Home Industries Section, died as did Chris Trautman, chairman of the Royal Holstein Club, and Alan van Wyk, the Society's longstanding public address announcer.

That year Judy Stuart was awarded honorary life membership after an association with the Royal Show that exceeded more than 40 years. During that time, she had served it in several capacities, especially in procuring overseas judges for the Cattle Section as well as extensive publicity. She was twice runner-up for the KwaZulu-Natal Female Farmer of the Year award and won the Female Master Dairymen's accolade. She had also contributed to youth development and had extended Iona Stewart's initiative of a Future Farmers competition into other provinces by establishing the Future Farmers Foundation.[58]

Physical improvements to the RAS Showgrounds and its buildings were quite substantial during the decade ending in 2011. Andrew Line, who served on the grounds and maintenance committee, recalled the prominent if not dominant role which Ron Glaister played for a long period in that regard.[59]

Improvements included the R800 000 worth of refurbishments to Hall 2 whose northern façade was later restored to its original specification in consultation with Amafa KwaZulu-Natal Heritage. Halls 3 and 4 were also subsequently renovated to become attractive venue functions outside Show periods with modern ablution facilities constructed in between them.

In addition, Halls 7 and 8 as well as the Crafts and Home Industries Hall were renovated, the roof of Hall 9 was replaced and hall exteriors painted and numbered. The installation of air conditioning in Hall 6 in 2006 meant that, in conjunction with the Olympia Hall, a full suite of comfortable facilities could be offered for plenary and breakaway conference sessions as well as dining venues. Adjacent roadways were widened to facilitate the transport of materials for major exhibits

The old clock tower was demolished, changes were effected to the front entrance extending from University (formerly Sappi) House to the Crafts and Industries Hall to make it more appealing and several entrance gates were replaced with 2 and 7 subsequently being increased by 30% capacity. The Sheep Section and clubhouse were upgraded as were the livestock wash bays to raise them to international standards, the cattle stalling and grooms' facilities were improved, a new loading ramp was constructed and Andrew Dickson sponsored a new livestock bridge over the flood control canal.

After the Kay Makan and Hirsch's buildings were relinquished, ownership reverted to the Society for other uses, as did the Stock Owners facility. The completion of Woodrite House adjacent to the Main Arena made a 50-seater conference venue available to the RAS for hire and also provided convenient storage for jumping equipment on its ground floor.

When the Pietermaritzburg Chamber of Commerce's sub-lease on Chamber House overlooking the Main Arena expired a new ten-year agreement was signed at R6 000 a month, subject to an annual escalation based on the CPIX.[60] It was decided not to spend R800 000 erecting a safety wall and catch-fencing around the Main Arena to create an oval track racing facility which was so popular elsewhere.

The residential property in Hyslop Road was redecorated while the recently acquired adjacent houses were demolished for incorporation into the Showgrounds. The former Natal Parks Board site was cleared to create

Terry's Green as an attractive stand for general exhibitions and in 2011 about 60 trees were planted to improve the aesthetic and eco-friendly appearance of the premises.

Importantly, water consumption was substantially reduced with the refurbishment of the reticulation system and in 2003 the municipality refunded the Society R60 000 for prior seepage and water loss. Further substantial savings were achieved that year by replacing single entry tickets with identity cards.

Cost-efficiency also induced the RAS to invest R140 000 in a 40KVA generator to obviate the loss of productivity caused by ESKOM's power outages. This adequately served the needs of the administrative offices, several halls and the main stage during show periods. In 2011 the Society invested in another cost-effective acquisition when it paid R10 000 for a cherry picker to facilitate tasks that involved working at heights. By then it was also belatedly realised that the RAS needed to address the shortcomings of its facilities in catering for visitors who were physically challenged.

Two heavy summer downpours in January and February 2008, both in excess of 75mm in less than 90 minutes, breached the flood control canal causing R55 000 worth of damage, mainly to the electrical reticulation system. Fortunately, this was covered by insurance but necessitated a careful re-assessment of the hydrology of the canal and of the stormwater drain emanating from Town Hill Hospital. This was re-laid at a cost of R205 000 to the Trust to complete the development of its Bank Park area.

In addition, a mini Dorpspruit wetland was constructed adjacent to University House which, it was hoped, would improve the quality of water entering the Showgrounds in conjunction with a second wetland to be created upstream in the Voortrekker High School grounds.[61]

Meanwhile parking facilities had come under increasing threat by business expansion in the vicinity of the Showgrounds. In 2003, 30% of the parking space on the polocrosse fields across Chatterton Road had been allocated to Barloworld for the relocation of showrooms and workshops with the prospect of a further 30% allocation, which the RAS vigorously but unsuccessfully opposed.

Efforts to negotiate a long-term agreement with Town Hill Hospital to provide parking space for 1 000 vehicles resulted in one hectare being secured for R6 000 for the 2004 Show. The mounting crisis for both the Royal and the Garden shows was thankfully eased by a further opening of the Voortrekker High School sports fields to the west of the Showgrounds, incorporating the old

Standard cricket ground and the hockey field beyond, for additional parking.

In addition, staff parking became available off Hyslop Road after recently acquired properties had been demolished. Traffic control was also becoming an increasing problem with the Society hoping to persuade the municipality to install appropriate traffic lights after spending more than R50 000 for the services of pointsmen during the 2011 Show.

The parking issue and need for traffic control was symptomatic of urban growth in the Showgrounds vicinity, across Chatterton Road and along Armitage and Sanctuary roads, with the prospect of more to come following the provincial government's confirmation of Pietermaritzburg as its capital city. These developments had ominous implications with regard to the imminent revaluation of properties and rates increases.

With these considerations in mind in 2006 the RAS approved a broad-brush master plan which was based on a 2001 plan and prepared by a special task team comprising Jan van der Vegte and David Hughes. This sought to provide for 'a mix of income-generating development' to ensure that the Society remained financially viable and did not appear to be obstructionist towards the necessary business expansion taking place around it. It was also intended 'at all times…to ensure the inviolate sanctity of the Royal Show and especially its agriculturally related activities'.

The plan envisaged splitting the Showgrounds into four primary components: a commercial section on Chatterton Road; a smaller office and possibly residential area; a four- or five-star hotel and exhibition, conference and convention centre; and, fourthly, 'an enhanced agricultural component', which was intended to remain its core business.[62]

It drew upon the expertise of architects, engineers and town planners and was intended to provide 'a directional guide' for the next decade if it attracted a positive response from both municipal and provincial government from whom appropriate rezoning might be required. It was well received by those two bodies. Assurances were given to all sections of the Show that no decision would be made to implement the plan without consultation with their committees and that any developments within the Showgrounds would be made via leasehold agreements.

The master plan was complete by June 2007 but was a dynamic, evolving document in which, among other things, a BEE partnership was briefly considered. A presidential succession plan was also devised to provide greater certainty in the future. Pleasingly, communication with the provincial Department of Agriculture was greatly improved which helped the Society's efforts to provide increasing support for the region's emerging farmers.

By 2009 the plan had been substantially diluted to eliminate those aspects 'which if followed through could have impacted negatively on the primary business of the Society'. It was still no more than a 'guiding document' for the future and included, *inter alia*, the agreed reincorporation of the Sheep Section into the Showgrounds from the property controlled by the Royal Mews Trust. In 2010 the master plan was put on display in the RAS conference room, along with a new clock presented by its longstanding bankers FNB.

Meanwhile there were ongoing discussions with prospective developers, municipal authorities and the provincial treasury concerning the construction of a conference centre and hotel. These proposals were prompted in part by the prospect of the provincial parliament being transferred to the Town Hill Hospital grounds. This did not eventuate and the proposals were dropped as not being 'in the best interests' of the Society or the Royal Mews Trust.[63]

Security also necessitated physical improvements. These included the erection of tower lighting and the completion of walling at the Chatterton Road stables and Victoria Siding parking. Petty theft and burglary within the Showgrounds continued to be an irritating distraction which could not be ignored, as was the discovery of counterfeit clip card tickets that required expensive counter measures. In 2003 all exhibitors were required to submit photographs for the preparation of new ID cards with which to facilitate multiple entry and eliminate ticket fraud.[64]

In 2009 all but one of the bronze otters, which Gordon Cunningham had cast and Tongaat Hulett donated, were stolen from the fountain at the entrance to the Olympia Hall. By then the theft of metal electrical and plumbing conduits had become prevalent, as was pilfering during the Society's major shows. This included the loss of the CEO's laptop on two occasions from his office. In 2011, 200 litres of diesel were stolen from the premises.[65]

Marketing was another aspect of future development that attracted increasing attention although expenditure on it increased from 9% to only 11% of revenue between 2001 and 2011. The adoption of a new Royal Show logo, featuring an ear of corn and the wheels of industry, helped to promote the Society's corporate identity.[66]

The return of the SABC to the Show in 2002 with an impressive interactive display which won an award the following year was accompanied by much more effective coverage on all radio and TV channels. The 2003 Show was particularly successful thanks to renewed efforts to raise the Society's profile through press releases while for R7 000 it secured a permanent one-page article in the monthly *Stock Owners News.* A survey of registration numbers in the car parks indicated that 60% of visitors emanated from the greater Durban area.

Independent Newspapers and the RAS shared the R400 000 expense of marketing the Garden and Leisure Show in 2004. This included large advertisements in *SA Gardening, Garden & Home* and *The Gardener* estimated to be worth as much as R30 000 of trade-in-kind. John Mienie of Berry's Pools contributed publicity as well as a pool to be won by a visitor.[67]

Unfortunately, the SABC's decision in 2008 to switch its marketing focus from exhibitions to educational mobile shows resulted in the Society incurring R300 000 in unbudgeted expenditure to provide alternative stage, lighting and sound equipment. However, following negotiations with SAMRO, a satisfactory agreement was reached concerning the music played over the Showgrounds public address system.

The already established RAS website continued to be successful and during the 2005 Show, for example, had 148 847 hits. In 2007 it was complemented by a new prospectus which provided a concise overview of the Society's history and activities. Importantly for both administrative and marketing efficiency its computer systems were upgraded periodically.

Telkom's omission of the RAS entry in the Midlands telephone directory on two occasions was a decided inconvenience as far as public relations was concerned although the installation of a new switchboard system in 2008 proved to be cost-efficient. Communication with the public at large was in good working order for the important events of 2011.

The **160th Anniversary of the RAS** that year was appropriately observed with fairly low-key celebrations considering the prevailing worldwide recessionary economic conditions and the consequent difficulties which many similar organisations were experiencing.

Although muted, the occasion did attract considerable attention on television, radio and in the press. The *Stud Breeder,* official publication of the SA Stud Book, even included a 60-page RAS celebratory supplement in its April 2011 edition. The numerous congratulatory messages received included greetings from the president of the RASC, HRH Princess Anne and from the provincial premier, Dr Zweli Mkhize, who was invited to open the anniversary Show.

In welcoming him the Society's president Iona Stewart observed that the proximity of the Showgrounds to the city centre greatly assisted the RAS to achieve one of its primary objectives: 'to introduce town to the country'. She reminded her audience that Dr Mkhize was himself a farmer, with a special interest in nGuni cattle. She also purposefully reminded him that when he had opened a recent nGuni Club sale at the Showgrounds he had stressed the agricultural importance of KwaZulu-Natal and had argued that 'more focus was needed on the agricultural potential of the province'.

CEO Terry Strachan stressed that the Royal Show was 'certainly the largest mixed agricultural show in Africa' in terms of livestock and equipment. There were more than 2 000 animals on site and exhibitors from as far afield as the Western Cape. They were among those attending the Hampshire Down, Hereford, Ile de France and Suffolk national championships during the 2011 Show.[68]

Appropriately, illuminated cattle figurines helped to create a festive atmosphere for the occasion and there was a reasonably balanced mix of agricultural, commercial, industrial and service sector exhibits as well as varied entertainments and other attractions. These included a succession of lively concerts as well as the contentious fireworks display, which was considered to be one of the most impressive ever seen in the province.

Attendance at such shows was in decline worldwide but the 146 312 visitors that the Show attracted was a remarkable improvement of 10 000 on the previous year. There was also much positive feedback from the public, some of which appeared in the media, while 83.7% of the 545 exhibitors ranked their experience of the Show above seven out of ten and only 1.4% responded with a score below five.

Following the successful achievement of that milestone there was every reason to face the future with confidence. Moreover, as Iona Stewart pointed out, the RAS could be grateful for the 'forward thinking' of previous executive committees. In the 1990s they had developed the Chatterton Road side of its property that now generated income without which its financial situation would not have been nearly as sound.[69]

She also cautioned that while the Society's 'credibility continues to strengthen both provincially and nationally' its activities were heavily dependent on the marketing budgets of its clients, some of which had been trimmed by as much as 70%. The interests of the RAS were 'predominantly show-related and as such, the vagaries relating thereto apply as much to us as they do to theatre and screen producers worldwide. Show-business is risk business and we simply cannot escape that reality'. Consequently, 'tight control on expenditure as well as exercising prudence in all our dealings' would continue to be the order of the day.[70]

In the years that immediately followed this indeed proved to be the case.

ENDNOTES

1 This chapter makes extensive use of *AR*s 2002–2012.
2 *AR* 2002: 10, 2009: 5.
3 *AR* 2003: 6, 2007: 14, 2009: 16.
4 Garth Ellis, interview, 13 September 2019.
5 Terry Strachan, interview, 12 March 2019.
6 RASM, Management, 18 June 2002: 2; General Manager's report, 18 June 2002: 1.
7 RASM, Executive, 23 June 2009: 5.
8 Iona Stewart Personal Papers, Stewart to *The Witness* (undated letter). Terry Strachan, interview, 12 March 2019. RASM, Agricultural Services, 22 July 2011: 2; Executive, 30 August 2011: 6.
9 Iona Stewart, interview, 12 February 2019. Terry Strachan, interview, 12 March 2019. RASM, Agricultural Implements, 1 February 2002: 2 and 19 April 2002: 1; Sheep and Goats, 29 November 2006: 1 and 20 February 2008: 1.
10 RASM, Executive, 29 January 2002: 5.
11 *AR* 2005: 6.
12 RASM, Sheep and Goats, 29 November 2006: 3, 18 July 2007: 1 and 20 February 2008: 3; Sheep, 3 November 2010: 2. *AR* 2011: 14.
13 Mike Moncur, interview, 26 June 2019.
14 ibid.
15 RASM, General Manager's report, 18 January 2008: 1. Guest, *A Fine Band of Farmers Are We!*: 224.
16 RASM, Poultry and Bird Representatives, 2 March 2002: 2.
17 RASM, Fur and Feather, 22 January 2009: 1.
18 *Farmer's Weekly* 8 July 2010. RASM, Executive, 20 April 2004: 3 and 24 April 2007: 4; Agricultural Services, 22 July 2011: 1.
19 *AR* 2007: 10. RASM, Management, 9 November 2005: 2.
20 *AR* 2007: 7. RASM, Management, 19 June 2007: 2; Executive, 26 June 2007: 3.
21 Iona Stewart, interview, 12 February 2019. Terry Strachan, interview, 6 March 2019. Reeva Wing, interview, 5 July 2019. RASM, Executive, 28 August 2007: 2 and 22 April 2008: 3; Management, 17 June 2009: 2.
22 RASM, Management, 21 June 2011: 2; Executive, 30 August 2011: 5.
23 Di Fitzsimons, interview, 4 June 2019.
24 ibid. *AR* 2003: 9.
25 *AR* 2008: 12. Di Fitzsimons, interview, 4 June 2019.
26 RASM, Management, 16 September 2008: 2.
27 Di Fitzsimons, interview, 4 June 2019.
28 ibid. Iona Stewart, interview, 12 February 2019.
29 *AR* 2007: 11.
30 Guest, *A Fine Band of Farmers Are We!*: 221.
31 RASM, Executive, 27 June 2006: 3.
32 *AR* 2003: 9; RASM, Garden and Leisure Show, 25 January 2002: 2, 21 June 2002: 2 and 19 July 2002: 2.
33 *AR* 2004: 9.
34 *AR* 2007: 12. RASM, Garden and Leisure Show, 20 October 2006: 1.
35 RASM, CEO's report, 18 November 2009: 4.
36 RASM, Garden and Leisure Show, 15 April 2011: 1, 11 August 2011: 1 and 28 October 2011: 1.

37 Andrew Line, interview, 29 July 2019. Gordon, *Natal's Royal Show*: 105, 131.

38 *AR* 2006: 5. RASM, Sheep and Goats, 23 March 2005: 4; General Manager's reports, 10 August 2005: 1 and 15 September 2006: 2; Cattle, 1 November 2005: 5; Executive, 27 January 2009: 4 and 20 July 2010: 5.

39 RASM, Management, 10 December 2003: 2 and 27 October 2010: 3; AGM, 27 October 2004: 3; Executive, 22 January 2008: 4.

40 RASM, Cattle, 21 June 2005: 6.

41 *AR* 2005: 5.

42 RASM, General Manager's reports, 11 February 2004: 2, 14 July 2004: 1 and 11 January 2006: 2; Management, 19 January 2008: 2.

43 Terry Strachan, interview, 12 March 2019. Andrew Line, interview, 29 July 2019. RASM, General Manager's reports, 26 November 2002: 1 and 27 January 2003: 2; Management, 21 February 2003: 1, 15 July 2003: 3 and 12 October 2005: 2; Executive, 20 April 2004: 5, 12 May 2004: 2, 27 October 2004: 3, 26 October 2005: 3 and 25 April 2006: 2.

44 RASM, Management, 15 July 2003: 1, 18 December 2003: 2 and 9 June 2004: 2; General Manager's report, 10 April 2007: 2.

45 RASM, Executive, 28 April 2009: 2.

46 John Fowler, interview, 10 June 2019. *AR* 2004: 6.

47 Iona Stewart Personal Papers, Speeches at Gold Cup Gala Dinner, 28 May 2007 and FNB Breakfast, May 2011.

48 RASM, General Manager's report, 21 January 2009: 2.

49 RASM, Executive, 25 June 2002: 3; Management, 12 October 2005: 2, 12 December 2005: 1 and 13 March 2007: 3. Shute, 'Royal Show': 34.

50 H.D. Spencer Personal Papers, Jenny Perry to RAS Executive, 30 August 2002.

51 Marlize du Preez, interview, 5 August 2019.

52 Kay Makan, interview, 2 July 2019. RASM, AGM, 23 October 2002: 5; Executive, 24 August 2004: 2 and 26 January 2010: 2; Management, 9 March 2005: 2, 16 September 2009: 2, 17 March 2010: 3, 17 August 2011: 3 and 14 September 2011: 2; General Manager's report, 13 July 2005: 2, 11 October 2006: 2 and 13 February 2007: 1–2; CEO's report, 18 October 2011: 4. *Witness*, 31 December 2018.

53 RASM, General Manager's report on RASC Conference, 20 August 2002: 6.

54 RASM, Management, 14 September 2005: 1–2, 9 November 2005: 1, 15 September 2006: 1 and 15 September 2006: 2.

55 RASM, Executive, 18 September 1997: 4, 28 August 2007: 5, 26 January 2010: 6–7 and 30 August 2011: 6; General Manager's report, 21 October 2008: 2 and 16 September 2009: T.D. Strachan's report on Midi Strategic City Summit, 29 October 2009: 1–3; Management, 16 February 2011: 1, 16 March 2011: 1 and 17 August 2011: 3.

56 Iona Stewart Personal Papers, 'Overview' (n.d.) and 'Beginning of it all'. Iona Stewart, interviews, 16 January and 12 February 2019.

57 Iona Stewart Personal Papers, 'Beginning of it all'. Iona Stewart, interview, 16 January 2019.

58 ibid. RASM, Executive, 23 April 2002: 4, 27 August 2002: 2, 28 January 2003: 2, 24 June 2003: 2, 27 June 2006: 5 and 25 August 2009: 6; Management, 15 October 2002: 4, 13 July 2005: 2, 19 February 2008: 3–4, 19 January 2011: 4 and 21 June 2011: 3; AGM, 22 October 2003: 2; CEO's report, 20 July 2011: 3 and 18 October 2011: 4.

59 Andrew Line, interview, 29 July 2019.

60 RASM, Management, 12 January 2005: 2, 19 February 2008: 2 and 19 August 2008: 1; Executive, 26 April 2005: 3.

61 RASM, General Manager's report, 19 February 2008: 2 and 18 February 2009: 2;

Management, 18 March 2008: 3, 3 December 2008: 1 and 19 October 2011: 3; CEO's report, 19 January 2011: 3 and 20 July 2011: 2.

62 *AR* 2006: 13. RASM, Executive, 20 August 2002: 2, 8 January 2004: 6, 24 August 2004: 2, 31 January 2006: 5 and Review of RAS Master Plan, 31 January 2006, 25 April 2006: 5; Management, 15 July 2003: 1, 14 October 2003: 2, 26 January 2004: 2–3, 15 April 2004: 1, 9 November 2005: 2, 12 December 2005: 1 and 21 June 2011: 1.

63 Andrew Line, interview, 29 July 2019. RASM, Management, 8 March 2006: 2, 9 October 2007: 2, 18 February 2009: 1 and 8 September 2010: 1–2; Executive, 24 April 2007: 5, 25 August 2009: 3, 26 January 2010: 2, 20 July 2010: 2–3 and 14 September 2010: 6; AGM, 28 October 2009: 4.

64 RASM, Cattle, 18 November 2003: 2.

65 RASM, CEO's report, 14 July 2010: 5 and 18 October 2011: 4; Management, 21 June 2011: 1.

66 RASM, Executive, 29 January 2002: 3.

67 RASM, Executive, 28 January 2003: 3; General Manager's report, 17 June 2003: 2; Garden and Leisure Show, 23 July 2004: 1–2.

68 *AR* 2011: 8. *Stud Breeder*, April 2011. Iona Stewart Personal Papers, Speech at May 2011 Show Opening. RASM, CEO's report on 2011 Royal Show, 21 June 2011: 1–2.

69 *AR* 2010: 16.

70 *AR* 2011: 22–23.

'CAUTION AND PRUDENCE' IN UNCERTAIN TIMES, 2012–2019

THE RECESSIONARY ECONOMIC CLIMATE that had characterised much of the first decade of the new century extended into the next.[1] As RAS president Iona Stewart observed in 2012:

> The business of hosting shows is largely dependent on the available marketing budget of one's potential client base and whether it is sensible or not, this is invariably the first cost centre to be trimmed (and even slashed) in difficult times. Similarly, the hosting of conferences and functions, whilst tending to be viewed by organisations as a pleasing optional add on to the activities of an entity, are seldom an imperative.[2]

Mike Moncur, who succeeded Iona Stewart in office, confirmed four years later that the 'exhibition based industry' still remained, 'at best tenuous' with 'limited client marketing budgets' playing a significant role in determining survival. In such circumstances 'the watch words caution and prudence' were as applicable as ever in managing the Society's business.[3]

Mike Moncur, vice-president since 2008, became president in 2013. Born and educated in Scotland he was an Underberg resident and a longstanding RAS member who was the immediate past chairman of the sheep and goat committee on which he had served with distinction since 2002.

His association with the Royal Show stretched back to the late 1970s when he was managing director of Drakensberg Gardens Hotel and in his spare time bred Dexters, which he exhibited at the Show. He also became interested in Hampshire Down sheep which he brought to the Show, along with budgerigars in collaboration with his father who had been an enthusiast for many years.

In the early 1990s Moncur left the hotel industry and sold his Dexter herd to concentrate on sheep but subsequently sold them as well after opening Mike's Restaurant in Underberg. He continued to exhibit budgies until his father's death in 2000. For a time, at the insistence of its outgoing chairman David Wing, he also assumed responsibility for the Horse Section.[4]

Marketing of the RAS and its activities became increasingly important. Only in this way could an ever-widening number of potential sponsors, exhibitors and members of the general public be made aware of what it had to offer by way of its annual shows and facilities for various other functions.

The question was raised as to why the word Natal in the Society's official name had never been updated to KwaZulu-Natal. In response it was argued that in 1904 royal assent had been awarded to what came to be known as the Royal Agricultural Society of Natal and that the subsequent name change of the province in 1994 was unrelated to it. Invariably it was referred to simply as the Royal Agricultural Society and so the matter was dropped.[5]

In 2012 yet another informative prospectus was produced, followed in 2013 by an illustrated leaflet for easier and less expensive distribution. The installation that year of a permanent WiFi zone covering the heavily used Olympia Hall and halls 6 to 9 was another useful marketing tool in attracting business. So, too, was the lease agreement entered into with Radio Hindvani for the use of the first two colonnades adjacent to Hall 2 for permanent broadcast and marketing purposes. This gave the RAS a year-round mouthpiece.

Its longstanding relationships with Maritzburg College and the City of Pietermaritzburg were strengthened by presenting both of them with framed certificates commemorating their 150th and 175th anniversaries respectively. Similarly, the presentation in 2013 of a framed centenary certificate to Malanseuns, the country's largest supplier of plant material, and to *The Gardener*, closely associated with the Garden Show, on its tenth anniversary was a constructive marketing and public relations exercise.[6]

The RAS also participated in the Meetings Africa Exhibition in Johannesburg which was useful for branding purposes. A further significant dimension of the Society's improved public relations involved much closer co-operation with the municipality following the appointment of Chris Ndlela as Msunduzi's mayor. An important consequence was the installation of a traffic light outside Gate 1 to assist pedestrians crossing to the Showgrounds after parking their vehicles on the Voortrekker School premises. Ndlela was to be missed when his term in office ended.

In 2014 the Society acquired another useful marketing tool when it was unexpectedly awarded a liquor licence, which had been refused some years previously. That year, following an earlier SWOT session, it also decided to spend approximately R1 million extending its traditional marketing campaign via print (primarily in Durban and Pietermaritzburg) and broadcasting to the electronic social media and a monthly website newsletter.[7] There were immediate positive results with the initial focus on the younger generation, employing the services of Easy2Access.

Various competitions were launched relating to the Royal and Garden shows using hash tags which offered opportunities to win prizes while also spreading information and growing the database. During the 2014 Show the Society's social pages (the website, Facebook, Twitter, Instagram and YouTube) were already among the most visited in South Africa. There were 20 049 searches, a reach of 540 363 people, 2 015 new likes, 13 573 people engaged and more than 30 000 hits on the website.

The RAS website and social media exposure was then further expanded and in 2015 the online marketing consultants reported that the reach on Facebook alone topped 1 200 000 while direct Facebook engagement with the Royal Show reached 43 822. Further, a video on YouTube proved highly informative in covering all aspects of the Show.

Newsclip reported that, in addition to paid advertising, the 2015 Show enjoyed R2.1 million worth of unsolicited exposure in the print and broadcast media. This increased to an estimated R6.3 million the following year.[8] In 2015, for the first time, provision was made for electronic Royal Show ticket sales via the internet. Only 292 were sold online rising to 437 in 2016 and 1 479 in 2017. It was another important start in a new direction.

Larger street pole signage was also deployed to provide better visual exposure in both Durban and Pietermaritzburg but by 2016 the marketing focus was very definitely on social media, supplemented by radio and print. This continued in subsequent years and was justified by website traffic amounting to 97 977 that year, as well as 11 448 page likes, 3 621 new Facebook likes and a Facebook reach of 2 500 593 which included not only the Royal Show page but also the greater social platform.

In 2017 the RAS website was redesigned and substantially updated, making it 'ultra-user friendly and easy to navigate'. There were more than 4 000 new likes, increasing the number of people connecting with the Royal Show to more than 15 000. Some 150 000 individuals visited the new website, 66 000 during the Show. In addition, various social media campaigns were conducted, reportedly 'yielding impressive reach and engagement statistics'.[9]

The Ticket Giveaway competitions achieved a reach of 551 082, the Royal Rock competition 152 240 and the Royal Show Laser Tag competition 32 239. By 2017, Royal Show South Africa was sixth on the list of Commonwealth agricultural shows on Facebook. The emphasis on digital and electronic media continued when website traffic increased to 218 524 in 2018 and 223 682 in 2019, supplemented by Facebook, Instagram and Twitter with Facebook impressions peaking at 685 536. In 2018 VOX Telecommunicatios became the Society's internet service provider.[10]

Three digital magazines and 28 videos were produced as vehicles for further promotions and, for the first time in years, the Royal Show enjoyed national television exposure on several channels following a trade exchange agreement with Multichoice. Although radio and hardcopy written media now seemed to be in decline, especially in reaching younger people, there was also substantial and still important coverage on East Coast Radio, Hindvani FM and RSG, as well as support from Independent Newspapers, Media 24, the *Farmer's Weekly*, *Landbouweekblad* and *Veeplaas*.

By 2018–2019 the increased marketing efforts were clearly broad-based and certainly provided much greater publicity for the RAS than a decade earlier. While it may have attracted much more attention from prospective exhibitors it did not seem to make any major impact on attendance and even less so on membership figures in the prevailing unfavourable economic climate.

Attendance at the Royal Show increased by 7 139 compared to the previous year to 153 451 in 2012 but that proved to be its peak for the next few years, declining to 124 664 by 2017 with other shows reporting as much as a 40% reduction in attendance. The average number of exhibitors had reached 460 by 2015, 54% of them industrial and commercial, 37% agriculture-related and 9% other, including crafts, home industries and government displays.

To make the Shows more attractive various inducements were introduced, such as Super Fridays which included lower gate charges, evening entertainment and special offers by caterers and the funfair. The first of these days, in 2017, resulted in a 24% increase in attendance; the busiest Friday in four years.

In 2015 approximately 50% of visitors to the Royal Show came from greater Durban, 41% from Pietermaritzburg and the Midlands and 9% from other provinces and abroad. Approximately 57% were between 18 and 40 years of age, 33% were older and 10% younger. Encouragingly, attendance rose again in 2018 to 129 050, doubtless assisted by ten days of good weather and national television and media coverage. These advantages did not prevent a subsequent 9.3% decline to 117 037 in 2019 although this was not as severe as that experienced at other shows. It was less than 47% of the record attendance set in 1998.[11]

The Garden and Leisure Show also continued to be vulnerable to the vagaries of the weather. Compared to 15 534 in 2011, the 2012 show attracted 20 152 visitors which would probably have been higher but for a cloudburst early on the Saturday when 50 mm fell in 90 minutes, effectively ending the day's activities. Fortunately, it was a four-day show that year to take advantage of the Heritage Day public holiday. So, too, was the 2017 show when attendance improved from 18 002 the previous year but still only stood at 18 125 and declined again to 15 595 in 2018.

Membership of the RAS, like attendance at its Shows, increased slightly in 2012 from 1 535 the previous year to 1 720 but, after peaking at 1 761 in 2014, it had dropped to 1 160 by 2018, despite R25 000 being spent on advertising in 2015 specifically to attract new members.[12]

This was attributed in part to the depressed state of the economy and to a declining interest in membership of associations and clubs being experienced worldwide. More recently there was also some resistance towards registering for the new biometric system implemented in 2017. Surprisingly, membership rose slightly to 1 266 in 2019 but still nowhere near the record 4 694 set in 1973. A computer program was implemented to monitor membership while fees continued, with occasional increases, to contribute roughly 1% of the Society's income.[13]

Main Arena displays and additional events at the Royal Show were still important in attracting attendance if not membership although, as former president John Fowler observed, packed grandstands were no longer a feature of the annual opening ceremony as before. However, by 2012 the small animal farm comprising donkeys, turkeys, geese and chickens provided a new interest for children while the chainsaw racing competition was rapidly becoming the national grand prix of its kind.[14]

There was the usual mix of entertainment and education. In 2012 the Main Arena featured the biannual performance of the SA Lipizzaners, on this occasion performing with the classical singing group Bella Voce on two consecutive nights in front of large audiences. Unfortunately, at a cost of R300 000 the Lipizzaners did not feature again but by 2015 the annual national dog agility championships seemed to have made the Royal Show their permanent home.

Captain Scully Levin's EQSTRA Flying Lions Harvard Aerobatic display gave five performances in 2012, including one in the evening with special lighting effects, returning the following year and again in 2018. Back on earth twenty motor vehicles demonstrated their own high-speed stunts across the Main Arena under the Sideways Drifter banner, followed in 2013 by the world-renowned Jungle Rush FMX bikers, by quad bike stunts in 2014 and monster truck demonstrations in 2015.

There was a military parade, a fireworks display and musical performances with KZN Premier Dr Zweli Mkhize as guest-of-honour in 2013 as part of a special opening ceremony to launch Pietermaritzburg/Msunduzi's 175th anniversary. Tragically that year the Royal Show was marred by the first death of a visitor in its 162-year history when six-year-old Simphiwe Mbense was killed apparently trying to extricate himself from a funfair ride while it was still in motion. Stricter safety measures were implemented to prevent anybody under 12-years-of-age from entering the area without adult supervision.[15]

The funfair, the largest in the province, had always been an attraction not only for children but also for adults and even members of staff. CEO Terry Strachan reportedly took 'great delight' in taking chosen employees and even executive members 'literally for a ride'. Iona Stewart recalled being treated to 'the Big Dipper and the Dodgem Cars' which necessitated briefly having to 'renege' on her presidential duties to recover from motion sickness.[16]

Each Show continued to include concerts catering for all tastes and ethnic groups, ranging from pop performances and RSG konserts to the KZN Philharmonic Orchestra which in 2012 featured a full choir and soloists to celebrate Queen Elizabeth's diamond jubilee. The concert series usually concluded with a now traditional Royal Show Bollywood music extravaganza and an Ukhozi FM concert which in 2012 attracted 22 367 devotees.

By 2014 declining attendance at the last event suggested that it may have reached its sell-by date as increasing numbers were enjoying the loud music while partying outside the grounds. In 2015 for the first time in sixteen years the Show ran for the full ten days with all exhibits open because that concert was not hosted on the final Sunday.[17]

The Vuma FM Royal Gospel Concert which replaced it unfortunately also drew disappointing audiences in subsequent years. In contrast the Hindvani FM Bollywood concerts, a mixture of comedy, dance and music, proved to be an ongoing success. So, too, were the annual KZN Philharmonic Orchestra's Royal symphony concerts, and the RSG performances which attracted as many as 5 000 fans. In 2018 the Royal uMgungundlovu Youth Festival, in collaboration with Gasasi FM and featuring prominent artists, disappointingly only attracted 1 000 fans.

The Working on Fire demonstrations held in 2013, 2016, 2017 and 2018 placed the emphasis on education rather than entertainment by impressing upon youngsters the benefits and dangers of combustion. The River Stage offered an interesting programme including shark dissections and talent shows, a 100-metre zip-line, plastic duck racing, fitness classes and a fun science secret lab.

In 2016 there was a break with tradition when the Show started with a three-day Enduro-X event using a track comprising various obstacles ranging from rocks, logs and sand to pallets, pipes and tyres. These all had to be rapidly removed in an all-night operation to prepare for the equestrian events. In yet another break with tradition the Cattle Gold Cup presentation, usually a closed event for breeders was, for marketing purposes, opened to the public for the first time in years and took place in the Main Arena in front of the members' grandstand.

In 2017 cattle, goats and sheep participated jointly for the first time in the Gold Cup parade and presentation. That year a Monster Activation Show was included, combining the skills of skateboarders, BMX and enduro bikers. There was also a Mad Max demonstration of car stunts as well as another Powasol Enduro-X event in the cattle arena after the livestock had departed. The 1 Special Service Battalion performed the retreat, a ceremony which it had initiated in 1934 and had come to be performed throughout the Commonwealth.[18]

A highlight of the 2018 Show was the arrival, for the first time in nearly ten years, of a train at the Victoria Siding with 160 passengers from Durban whose excursion was largely completed under steam power. Another unusual feature was the Red Meat Producers Organisation (RPO) sponsored 'Brekfis met Derrich' which Derrich Gardner produced for RSG and 120 invitees while reaching an estimated listening audience of 200 000.

The new Soweto Towers Bungee and Scad Freefall 50-metre jump failed to attract the public, possibly due to the cost. Crowd-pullers included the Coca-Cola-sponsored Monster Freestyle FMX Night Fight involving a mixture of pyrotechnics and dexterous professional motor-cross riders, the Royal Demolition Derby stock car race and yet another Powasol Enduro-X performance in the cattle arena. This took place after all the animals had departed and with lower noise levels in response to earlier criticism.

The livestock, as ever, remained all-important for while every Royal Show sought to present a balance of agricultural, commercial and social service exhibits as well as spectacular entertainment, the traditional emphasis was still on the first of these.

The **Agricultural sections** therefore continued to constitute the core of the annual Royal Show despite the other crowd-pulling attractions. In 2012 the RAS decided to follow international trends by entrusting a large part of the administration of the livestock sections to the breed societies thereby easing the burden on its own staff.

In all 1 352 animals were exhibited that year but there was growing concern that, in the prevailing economic climate, many farmers might be obliged to review the cost of participating in agricultural shows. It was anticipated that if in future some financial assistance became necessary the Society might achieve this by cross-subsidisation from its other more profitable activities.

A welcome change in 2012, after a 30-year absence, was the re-introduction into the Show of a seed and vegetable component. This, it was anticipated, would grow into something much more substantial, like the giant pumpkin seeds which the RAS itself had sown. A giant pumpkin competition was included in the theme 'Follow the Seed'. The intention was to illustrate the vital role which tiny seeds played when transformed into food, medicinal products and clothing. Unfortunately, in 2013 the South African National Seed Organisation withdrew its participation.[19] The livestock sections continued as before.

The **Cattle Section** experienced a slight drop in entries in 2012 but successfully hosted both the Suffolk and Dairy Swiss national championships. Exhibitors were drawn from as far afield as Vryburg and Franschhoek, re-affirming that the Royal Show had 'evolved into a show of national importance'.[20] For the second year the Cattle Expo enabled all the breed societies to disseminate instructive information among visitors.

In 2013 the focus was on the youth and Future Farmers as well as commercial on-the-hoof judging with 50 entrants. There was a re-emphasis on the need for young exhibitors to have fed and trained their animals instead of simply borrowing them for competition purposes.[21] Overall winners were the Owen Sithole College of Agriculture, which secured R5 000 in prize money.

Classes of non-registered heifers were introduced to broaden the scope of the section while Angus and Suffolk national championships were again held but there was a slight decline in dairy category entries. For the third time in twenty years a cow broke loose from the stalls causing some injuries to a panicky public, the worst being a broken arm. Three cattle restraining barriers were subsequently installed along the relevant roadway.

Sadly, that year Joan Crouse, who had been the backbone of the Youth Show for many years, resigned from the committee but a replacement was found in Terri Chubb, assisted by Di Oldfield. Happily, a successful Youth Show was held in 2014 with more children becoming involved and a team of 24 being chosen to represent the province at the nationals in Vryburg.[22] The Cattle Section suffered a further loss when its chair Grant Kobus, who had been actively involved for many years, emigrated to New Zealand.

In 2014 entries declined again slightly to 504 but the Cattle Expo continued to be one of the biggest events of its kind. It drew exhibitors from as far distant as 1 000 kilometres and attracted public interest, not least to its on-site milking parlour. The quality of specimens on show was still excellent, particularly in the heifer category, although there was a worrying decline in cattle entries due to increasing costs and labour issues. The supreme champion beef bull and beef cow, Bertie van Zyl's PinZ²yl and C.A. Froneman's Simbra, both earned special accolades from the judges as did Neil van Rensburg's supreme champion Holstein dairy cow.

In 2015 the 530 entries exceeded expectations in the prevailing economic and drought conditions while the Ile de France national championships that year introduced several new exhibitors to the Royal Show. Unhaltered trained livestock (Simmentalers) were entered into the stud category for the first time, as was a commercial competition and auction for weaners that was intended to provide commercial farmers with more show exposure.

As always, the primary focus was on the stud animals with the Angus, Brahman, Braunvieh, Hereford, Holstein, Jersey, Limousin, Pinzgauer, Simbra, Simmentaler and Sussex breeds all represented. That year University of KwaZulu-Natal made a welcome return after six years to the open and the student challenge in competition with the agricultural colleges. The Show was marred by bad behavior on the part of unsupervised school groups on cattle judging days. In response, barriers and security guards were installed at the northern and southern ends of the arena and Dr Sishi, Head of the KwaZulu-Natal Education Department, conveyed the Society's complaint to all provincial schools.[23]

In 2016, when there was a major upgrading of the cattle handling facilities, the 498 entries again exceeded expectations. Heifers and weaners competed for the second consecutive year and a wide range of breeds was again represented with most of the 121 steer entries emanating from the agricultural colleges.

Another Youth Show and Future Farmers exhibition was organised to encourage and educate the younger generation, the former attracting 108 entrants. Vryheid Landbouskool returned to the Show after a long absence and won the champion on the hoof award and that for first reserve in the carcass competition, which augured well for the country's agricultural future.

In 2016 Iona Stewart announced her retirement after 45 years on the cattle committee and was thanked for her 'enormous contribution'. Following the decline in Hereford and Sussex entries at the Show that year it was subsequently decided, as a pilot project, to allow judging to take place on the farms in accordance with the standards of their respective breed societies. The champions would then be eligible to participate in the Gold Cup competition at the Royal Show.[24]

Pleasingly, entries in all the livestock sections increased in 2017, from 498 to 504 in the case of cattle with Dexters making a welcome return after a long absence with seventeen specimens. In response to repeated requests from farmers, the cattle and sheep programmes were shortened from seven to six days. Among other expenses that exhibitors incurred was the high cost of local bed and breakfast establishments.

By then it was being suggested that to avoid dead zones in the Showgrounds the use of cattle clubhouses should be optimised during Show periods. If necessary, they could be merged into just two, serving beef and dairy exhibitors respectively, while the others could be demolished or re-allocated. The breed clubs were subsequently advised that they were obliged to be fully operational from the opening Friday up to and including the first Wednesday of each Show, that rentals should be paid by due date and that non-compliance would result in the loss of occupation rights to their clubhouses.[25]

Meanwhile the Cattle Expo doubled in size, enjoying strong support in 2017 and again in 2018. The highly prized Wagyu breed was on display, among several others, as well as the well-known RPO painted cow, which served to illustrate the location of various meat cuts. Following their absence the previous year, there were a substantial number of dairy cattle at the 2018 Show, thanks to strong support from the KZN Jersey Club and sponsorship from Agri. Livestock entries increased again overall from 3 694 to 4 069. There were 662 cattle entries, including 500 stud cattle, with most breeds well represented but, as usual, Brahman were the most numerous followed by Simmentaler and Simbra.

PMB Petroleum generously sponsored 600 litres of diesel to the section for the use of exhibitors that year. This greatly assisted a Vryburg farmers' syndicate, the Vryburg Beef Study Group, to enter 50 steers in the commercial category after being attracted by the favourable prices at the Show. The heifer classes also increased in numbers and several 2018 exhibitors declared that the Royal Show was still 'the preferred South African livestock show'.[26]

Prominent among the many award winners during this period were Brandwater Brahmane (RAS Gold Cup beef bull supreme champion in 2012 and 2018, beef cow supreme champion 2017 and Meadow Feeds Trophy beef cow reserve supreme champion 2012), P. Bester (beef bull supreme champion and reserve supreme champion in 2016), Kevin Lang (RAS Gold Cup dairy supreme champion in 2012 and 2017, John Simpson Memorial Trophy dairy reserve supreme champion in 2015) and J.A. Theron (RAS Gold Cup dairy supreme champion 2013 and 2016 and John Simpson Memorial Trophy in 2016 and 2017).

Angie Malherbe from Mid-Illovo made history in 2017 when she became the first female breeder to win the RAS Gold Cup for the supreme beef bull on show, a Limousin named La Rhone Millionaire. Some years previously Judy Stuart had won the dairy cow Gold Cup.[27]

The **Sheep/Lamb/Goats Section** also continued to thrive. Yet again it provided a venue for the Hampshire Down national championships in 2012 and 2013. It is uncertain to what extent the substantial increases in both sheep and goat entries were attributable to the Facebook page created to increase interest in that section. This was particularly pleasing considering that the province was not a major goat producing region and served to emphasise the countrywide recognition of the Show's importance. The Sheep and Natural Fibre Expo continued to attract public interest with fashion shows, sheep shearing demonstrations, interactive displays and natural fibre products.

In 2014 the Show hosted the Border Leicester and Hampshire Down national championships as well as witnessing the return after a long absence of SA Mutton Merino. Next Generation volunteer in the Sheep Section, Danie Raath, made such an impression that he was offered a prominent managerial position in KwaZulu-Natal. The RAS recorded it as 'a clear indication that the opportunities and synergies that exist between the Society and the Next Generation can translate into positive benefits.'[28]

The following year there were 620 sheep and goat entries, one of the largest in recent years, including Boer Goats, Border Leicesters, Dorpers, South African Mutton Merinos and Suffolks as well as entrants in the Hampshire Down and Ile de France national championships. The Agricultural Research Council also participated in the ever-popular expo with a display of indigenous small stock.

In 2016 the Show again hosted the Hampshire Down nationals as well as the Boer Goat regional championships and, for the first time, SA Kalahari Reds participated. This was welcomed as a further 'vindication of the Show's relevance beyond the borders of KwaZulu-Natal'.[29]

The 473 entries in the Sheep, Lamb and Goat Section included 160 goats, 90 of which comprised Jan Hepburn's single entry of Boerbok and SA Kalahari from the Waterberg district in Limpopo. He and Albert van Zyl of Soetdorings each successfully entered a champion and reserve champion in the supreme ewe and ram categories respectively.

Entries increased marginally in 2017 from 473 to 558, boosted by 101 Ile de France, brought there for the national championships. As before, the expo continued to attract special attention while the KwaZulu-Natal leg of the Toyota Young Auctioneers Competition was held at the Show for the first time. Gavin Matthews from Estcourt qualified to advance to the November finals at the Alfa Show in Parys where he won the senior division. An informative one-day SABC Education Living Land Workshop was also held for 200 emerging farmers in partnership with the Department of Rural Development and Land Reform.

Entries were up again in 2018 to 681, comprising 244 sheep, 144 fat lambs and 293 goats. The Hampshire Down national championship was held for the 15th consecutive year and an indigenous goat sale was introduced, thereby compensating for the dearth of livestock during the final weekend of the Show. Matthew Fyvie and Fiso Hadebe, the latter from Weston Agricultural College, won the senior and junior categories respectively of the Toyota Young Auctioneers Competition.

Apart from its 'traditional displays and activities' the Royal Show was by now 'increasingly being used to more formally afford training and disseminate information'.[30] The day-long Living Land Workshop, also first introduced the previous year, was repeated with specialists presenting thirteen instructional modules to 200 emerging farmers thanks to sponsorship from the Land Bank and SABC.

In addition, with the sponsorship of the RAS and Plaas Media, Woodrite House provided the venue for a full-day course on intensive sheep farming which more than 60 commercial and emerging farmers attended. The Sheep and Natural Fibre Expo was given a major make-over in 2018 with strategically placed wooden pallets replacing the previous exhibition shell scheme. It continued, as before, to attract attention with its informative and varied mix of exhibits, food, livestock and interactive demonstrations.[31]

Among the winners of the two trophies available in this section were Twee Seuns Suffolk who won the William Cooper & Nephews Floating Trophy/ Farmer Floating Trophy for the supreme champion ram in 2012 and 2017. Rooihek Dormers won the Epol/William Cooper Floating Trophy for the supreme champion ewe in 2014 and again in 2015 and 2017, as well as exhibiting the supreme champion ram in 2015.

The **Carcass competitions** also went from strength to strength with the

2012 Show itself attracting three awards at the SAMIC carcass competition. Apart from the award winners, which exceeded the ruling prices by up to 400%, those in the sheep category were disappointing, especially for lamb. The goat category was a new addition, as was the competition for lamb breeds with substantial European genetics which had a slightly higher fat content.

In 2012 G.J. Lotter Jnr of Hofmeyr emulated the earlier achievements of his father by exhibiting the champion and reserve champion in the lamb category, both White Dorpers, which scored 98.79% and 98.46% and sold for R240 and R250 per kg respectively. J.J. van der Colff of Williston won the award for the champion group, also White Dorpers, which sold for R60 per kg.

A.J. van Heerden of Bethlehem bred the champion lamb with substantial European genetics, a Hampshire Down scoring 94.75% and selling for R70 per kg, while G.J. Hoon of Williston's Boergoat was the goat champion, scoring 91.91% and selling for R168 per kg. Following objections, goats were not included in the 2013 competition. No records were broken, although the R475 per kg for Gert Lotter Snr's winning White Dorper lamb carcass was the best price in the country that year.

In the beef carcass category Weston Agricultural College produced the champion in 2012, a Bonsmara which scored 96.16% and sold for R68 per kg, as well as the champion group, also Bonsmaras. In 2013 a new South African record price of R110 per kg for beef was established by R.H. Köhne's winning Simbra X Braford entry at a time when the ruling national price was R27 per kg. It was interpreted as yet another indication of the 'perceived prestige of the Royal Show carcass competition'.[32]

In 2014 Owen Sithole Agricultural College produced the champion beef carcass, which scored 97.88%, and the champion group, while Cedara Agricultural College produced the first reserve champion, all Bovelders. There were more than 200 buyers at the auction and, for the first time, a pig carcass competition was included after Mike Moncur had met the Natal Pork Producers' Organisation. Breeders were more than willing to participate, in contrast to their reluctance to bring live animals to the Show for fear of infection. Joseph Baynes Estate produced both the champion and first reserve carcass.[33]

This made the Royal Show the only show in the country that included the three species in its carcass competitions. In 2015 it received a certificate, trophy and prize money from SAMIC for achieving the highest group percentage of 93.10% for commercial pork, as well as a certificate acknowledging it as the largest agricultural show in the country with a national carcass competition.

The 2013 record was broken when a national record price was set for beef with the Oyster Box Hotel paying R150 per kg for UKZN's Bovelder, which achieved a 97.24% score. That year the second highest price ever achieved in South Africa for lamb was also set at R1 100 per kg by RCH Ile de France's champion entry, which scored 98.32%. This was bettered in 2016 by the R1 600 per kg and 99.24% achieved by Coetzee Reitz's White Dorper, although neither matched the R3 000 per kg world record set at the 2007 Show.

By 2016, with 107 cattle, 144 lambs and 80 pigs entered, the carcass competition was in its 25th year and firmly retaining its reputation as the country's premier annual red meat event. Craig Peters was well established as its auctioneer and all cattle and lamb slaughtering was still being undertaken at Jacques Abattoir in Estcourt in accordance with strict Halaal procedures. Thereafter the carcasses were displayed in a cold room that Booysens Refrigeration sponsored and maintained.

In 2016 Glenelly-Craig McCord's champion pig set a new national record of R95 per kg. Regular bidder the Oyster Box Hotel in Umhlanga bought all three champion carcasses that year. The number of buyers at the 2017 competition declined from 25 to fourteen and bidding was disappointing with no new record prices.

The Mtubatuba Mini Farm educational project did exceptionally well in the beef competition. It produced the champion, a Bonsmara, which scored 97.78% but was sold to the Oyster Box for a disappointing R70 per kg, as well as exhibiting the reserve champion group. It was the third time in recent years that an educational institution won a championship auguring well for future livestock production.

Regina Harmse from the Ermelo district exhibited the champion lamb, an Ile de France with a score of 99.37%. It was also the European Genetics champion and was sold to the Oyster Box for R1 350 per kg. At the request of breeders, super lamb and super ox categories were introduced to acknowledge higher weight excellence above 50 kgs and 315 kgs respectively. T. Steenkamp's Suffolk X Dormer won the former and Vryheid Landbouskool the latter with the Oyster Box purchasing both carcasses.

That hotel also scooped the champion carcasses in the pork competition which, for the first time, catered for emerging farmers who had been in the breeding business for not more than three years. N.F. Mbalo produced the champion and reserve champion in that category, both PICs, the former being sold to the Oyster Box for R26 per kg. Gerd Baum of Joseph Baynes Estate exhibited the overall pork champion, also a PIC, for which the Oyster Box paid R30 per kg.

By 2018 what was now called the Landbouweekblad Royal Show Carcass Competition was in its 27th year and attracting entries from as far afield as Barkly East and Williston. On that occasion the Vryburg Beef Study Group did not break any records with their Bonsmara but 'amazingly walked away with every major accolade in the bovine section; the first occasion in 22 years where a single breeding entity achieved this result'.[34]

It included the super ox competition which, with that for super lamb, was again included at the request of breeders. The latter was won by Twee Seuns with a Suffolk entry but former multiple winner Gert Lotter's White Dorpers won the champion lamb and champion group titles. NHELC Farm produced the champion PIC pig, Butt Farming the champion Topig group and S. Mlangeni the champion Landrace in the carcass competition for emerging farmers.

Bids among the 200 plus audience were fairly good for lamb and pork, but disappointing for beef. Kevin Joseph, executive chef at the Oyster Box, was again prominent in securing several of the best carcasses in all categories. Following a meeting with relevant role players it was decided to continue with the disappointing beef section in the expectation that it would improve by way of national bidding and sponsored TV coverage.[35]

There was no such concern about the lamb and pork categories. Other sections of the Royal Show were similarly highly competitive.

The **Bird and Poultry Section** held all of its major exhibits at the Showgrounds but several of them took place outside Royal Show time. They nevertheless added an important dimension to each Show and continued to attract public interest.

The 2012 Show included a closed club display which the Pietermaritzburg Canary and Cage Bird Club mounted. It was an impressive exhibition but its exclusivity did result in a reduced number of entries. The 101st annual open show which the club held the following year during the Show attracted more than 950 exhibits, many of them from outside KwaZulu-Natal, while membership increased by 20%.

A closed club display was again held during the 2014 Show but, as before, the club's major events took place outside Show time, most notably its 102nd annual show in June and young bird show in May. The following year the winter provincial exhibition was held during the Royal Show attracting 44 exhibitors and 739 birds. There were another 730 entries at the 2016 Show and as many as 943 in 2018, some from the Free State for the first time, when the provincial competition was again held during the Show.

The Pietermaritzburg Budgerigar Club, founded in 1958, sold 55 birds at the 2012 Royal Show but its biggest event of the year was the one-day display organised at the same venue outside Show time on behalf of the Budgerigar Society of South Africa, which attracted 400 entries from all over the country.

In 2013 the Canary Club and the Budgerigar Club shared the Bird Hall. The latter helped to increase the overall number of exhibits to more than 1 200 by staging a provincial competition during Show time for the national Budgerigar Society. It did so again in 2014, but outside Show time, while its own annual club show was held during the Show and was therefore much smaller.

In 2015 and again in 2016 and 2017 the Budgerigar Club's biggest event was a two-day provincial show on behalf of the Association for Wild-Type and Exhibition Budgies of South Africa which attracted nearly 200 entries in each of those years, some from far afield. Only the last of these was held during the Show, as was the provincial show held in 2018, again with 200 entries.

The **SA Fancy Pigeon Association** was also well represented in 2013, when its regional exhibition was outside Show time, and again in 2014. By then the further growth of the Bird Section as a whole was constrained by shortage of space with each species being allowed only limited display time. In 2016 there were 2 501 entries altogether.

That year the Fancy Pigeon Association alone attracted 750 high-quality entries representing 53 breeds to the Royal Show with Craig Pitt from Underberg exhibiting the best bird on show, a Medium Crested Helmet. Entries were well down the following year with reports of an avian flu outbreak but the club continued to encourage and mentor its junior members. At the 2018 Show there were 435 birds on display, representing most South African breeds but all from only four exhibitors.

Poultry attracted more than 800 entries in 2012 and again in 2013 for the third consecutive year. Although the 50-member strong Natal and Coast Poultry Club by then constituted the third largest in the country, many of the 40 exhibitors were from outside KwaZulu-Natal and in 2013 six were first-timers. In 2014 there were 940 entries, the largest in fifteen years, and 43 exhibitors, the most in 30 years.

That year and again in 2015 there were a pleasing number of first-timers participating. Entries peaked at 900 in 2016 with hectic sales reflecting increased public interest. There were 705 entries the following year and 709 in 2018 with a noticeable increase in the number of juniors. Interest in the larger breeds was growing with Wyandottes and duck exhibits gaining in popularity.

The **Honey Section,** as the apiarian exhibit was now commonly known,

Well-known poultry breeder **Tim Nixon** began a long term as RAS vice-president in 2013. He had been exhibiting Rhode Island Reds at the Show since 1972 when he was five years old and joined the Natal and Coast Poultry Club. He followed in the footsteps of his grandfather who participated in the Show for 40 years from the late 1920s while his father was a vet who specialised in poultry diagnostics. Tim Nixon's interest in poultry subsequently extended to other breeds, but it was strictly a hobby conducted on his Birnam Wood smallholding as he was engaged professionally in the animal feed business. He also kept cattle in the Free State which he exhibited on the Show.

In 1984 he was invited onto the committee of the Natal and Coast Poultry Club which he served in various capacities and in 2004 began to show rabbits as well, becoming chairman of the Natal Rabbit Club the following year. From 2009 he was chairman of the feather and fur (birds and rabbits) committee which was officially recognised by the RAS executive. This, coupled with his subsequent election as vice-president, gave greater prominence and a stronger voice to those sections than they had hitherto enjoyed.[36]

was another regular feature of the Show and continued to gather strength under Craig Campbell's chairmanship. It enjoyed an increase in entries in 2012 and attracted public interest with its ongoing dissemination of information about the essential contribution bees make to the broader environment. In response to urgently needed repairs to the Honey Hall the RAS undertook to deal with the structural damage caused mainly by roof leaks while the KwaZulu-Natal Bee Farmers Association, custodians of the venue, revamped the interior.

By 2013 the association was attempting to expand the annual event into a national honey exhibition but there were few participants from outside the province and other regional beekeeping bodies were slow to embrace the idea. Membership of the local association was increasing and with it by 2014 the number of entries was the highest in a decade.

Unfortunately, sales, upon which the association largely depended for income, were still lagging due to the prevailing economic conditions. Members of the South African Bee Industry Organisation and the Southern African Apiculture Federation were suitably impressed when they visited the Honey Hall and undertook to encourage beekeepers in other regions to support the Royal Show.

The number of entries in 2015 was one of the highest in recent years, including several from other provinces, while there was also an increase in the sale of honey and related products. Conveniently, that year the Show coincided with the National Bee Keeping Congress, which attracted 120 delegates from all over the country. Entry levels and public interest were maintained in 2016 with several exhibitors emanating from outside KwaZulu-Natal and a citrus-based honey from the Western Cape being declared the champion.

There was a record entry of 600 bottles on display the following year but only one from outside the province which was a reflection of the current economic climate and the drought in the Western Cape. Sarah de Silva produced the champion bottle while Steve Smith dominated the novice category. The Honey Hall, which also disseminated information about adulterated products, or honey fraud, won the Pietermaritzburg Chamber of Commerce trophy for Best Small Business Exhibit.[37]

The campaign against fraud was continued at the 2018 Show when the Honey Hall was adjudged Best Display in an Exhibition Hall for which it was awarded a gold medallion and the Royal Commercial Floating Trophy. Entries and sales were slightly down, again with no participants from the Western Cape. Craig Campbell (Peels Honey) and Sheila McCall (Hawthorne Hill Farm) emerged as the champions.

The **Rabbit Section** was still well represented at the Show, attracting nearly 200 entries in 2012 and for the first time filling the remaining cages with a display of guinea pigs. Exhibitions in subsequent years were similarly successful as was the sale of rabbits and rabbit products. In 2014 there were more than 300 entries and 22 exhibitors while club membership was also growing, with the assistance of the RAS website. This upward trend continued in 2015 with judging having to take place over two days as it did again in 2016 when there were 412 entries.

For the first time the section was over-subscribed in 2017 with 432 entries and 27 exhibitors, necessitating the construction of 100 extra cages at short notice. Public interest and sales were as strong as ever with entries increasing marginally to 439 in 2018. Weston Agricultural College pupils exhibited for the first time that year as the annual Youth Show was extended from the cattle and sheep sections. International judge Derek Medlock, chairman of the British Rabbit Council standards committee, attested to the quality of the animals on display.

At every Royal Show there were the usual alternatives for those who were attracted by much larger exhibits such as machinery.

Agricultural machinery was initially much in evidence with most of the tractor and implement manufacturers participating. In 2012 every available stand was occupied. Mascor relinquished some of its space for a new exhibitor, but several others could not be accommodated. By 2014 even the additional 3 000m² now available for agricultural and other machinery adjacent to Hall 9 was oversubscribed with good sales reported. However, in 2016 agricultural machinery sales dropped by 28%, which was drought-related although some major participants did not attend. By 2018 agricultural equipment and tractor manufacturers were again much better represented.[38]

For those who preferred large livestock, horses among others were still prominent at the Show.

The **Equestrian Section**, although seemingly no longer such an integral part of the Royal Show, continued to go from strength to strength under Colin Scott's capable co-ordination. By 2012 the Main Arena had at last achieved full lawn coverage following the BMX championships held there in 2010 and 30m³ of sifted river sand was applied to it. The Show involved no less than three jumping competitions, including the KZN Adult open championships, the SA Pony Rider Outdoor grand prix and the SA Junior Outdoor grand prix. The standard and number of entries was as good as ever, comparing favourably with events elsewhere in the country.

Sue Cullen's imported stallion Wenzel R was the supreme champion that year with Robyn Bechard's Dimmock Dragonaire winning the Gold Cup for champion hack. This was shown by Robyn Berry who was awarded the Victor Ludorum. Stephanie McGaw's Waterside Fair Exchange emerged amid stiff competition as the supreme champion child's pony while the double slalom, tent pegging and other competitions provided further attractions.

In 2013, under the guidance of Colin Scott, Jacquie Pappas and Priscilla Young, the equestrian programme concentrated on young riders and a national jumping championship for ponies before concluding with a two-day exhibition tent pegging competition involving the national and provincial teams.

Despite the generally good condition of the Main Arena's surface, several contestants reported that it was still too hard. By then the rising expectations of internationally competitive riders was making it increasingly unlikely that it would continue to meet their requirements unless the venue was developed exclusively for equestrian events. Ideally, in view of the enormous cost of purchasing and maintaining quality horses, it would necessitate the installation of a modern synthetic surface. From the Society's perspective this was impractical in view of 'the eclectic mix of Royal Show activities'. In future the focus at the Royal Show was therefore likely to remain on young riders.[39]

Significant changes were then taking place in the equestrian field. In response to the Department of Sport and Recreation's requirement that, in common with other sports, equestrian activities should be restructured along club lines, the RAS inaugurated the Royal Riding Club. Based at the Showgrounds with an initial 36 members, its purpose was to give potential riders in its region a means of entering the sport as did the other new clubs in the province: Shongweni, North Coast and Midlands (north of Howick).

Following the amalgamation of the Midlands clubs the Royal Riding Club was subsequently disbanded. So, too, was the 61-year-old KwaZulu-Natal Horse Society along with other regional horse organisations, in accordance with the new restructuring requirements. It was then resolved to accommodate its considerable memorabilia in a Museum of the Sport Horse in the downstairs office of the equestrian clubhouse for public display.[40]

There was still a busy equestrian programme at the 2014 Show, including a combined adult and children's class, a double slalom knockout competition, demonstration polo matches and a tent pegging competition between the Jordanian and South African women's teams, all of which was intended to attract the public. For those with more serious equestrian interests Ireland's Jane Bradbury joined other out-of-province judges for the showing classes. Among the various breeds represented a whole day was devoted to Arabs.

The Main Arena was no longer considered suitable for A-grade jumping which therefore moved to specifically prepared venues, but the Showgrounds were still regarded as the best available countrywide for the showing classes. In 2015 Australian Annette Vickery was invited to judge a good turnout of general breed and Arab showing classes while another medley of competitions added variety to the occasion.

That year there was further confusion among the equestrian fraternity following divisions within its national federation which created uncertainty as to whether or not membership of a riding club was indeed compulsory. In anticipation of this the equestrian clubhouse was improved and the sand warm-up arena enlarged and resurfaced. The Royal Riding Club was eventually reinstated following this 'period of turmoil'.[41]

At the 2016 Show entrants from as distant as the Northern Cape competed in another medley of events. They included dressage, showing, jumping, tent pegging, carriage driving and, for the first time, the SA Boerperd regional championships. This involved competitions and classes not often seen in the province. Space and time constraints in the Main Arena due to the preceding three-day Enduro-X competition made it impossible to hold the showing classes concurrently with the Royal Show that year, even though it was still regarded as the best available venue for such events.

These classes were again included in 2017 with international judge Lesley Whitehall adjudicating and further enhancing the credibility of the occasion as the Royal Show sought to consolidate its position as the country's leading equestrian showing venue. The rest of the programme included a familiar mix with 30 competitors, many from other provinces, competing in the Boerperd events which now appeared to have become a permanent feature. The last three days focused on another international tent pegging competition between the Proteas women's team and the Jordanians.

Again, there was a full and varied programme in 2018 featuring, pre-eminently, eleven showing classes and championships (with entries increasing from 80 to 160 in a year), 21 dressage classes and championships, eight show jumping classes, 51 riding classes and championships, the Boerperd championships and mounted games. British judges Lesley Whitehall and Terry Chalmers were impressed with the overall quality and considered the showing classes to be of international standard. There were generous sponsored prizes with Carol Aronson's impressive ex-racehorse South Lawn and its trainer/rider J.J. Kemp winning seven championships.

The **Crafts and Home Industries Section**, as ever, produced its own brand of champions in presenting a wider variety of domestically made items than any similar showcase in South Africa. Individuals, organisations and schools continued to compete, with the photographic entries in particular maintaining their usual standard of excellence. In 2012 there were 156 new entrants and 450 additional exhibits. Embroidery and knitting were particularly well supported, generating an urgent need for more display cabinets.[42]

Entries were on a par the following year with Leon Hayes winning the Father Alston Trophy in the highly competitive photography category and Dr Rob Hart securing the most points in 2014. By then there were nearly 500 photographic exhibits, 94 quilts and large numbers of high-quality entries in the embroidery, knitting and needlework categories. Indicative of fluctuating interests, knitting entries increased by 90% in 2015 while those in other categories declined but the standard of woodwork items improved markedly and that of photographic submissions was as high as ever.

There were 3 231 entries and 287 exhibitors in 2016 with the emphasis 'on interactive and informative demonstrations' involving 'lace making, weaving, tatting, leather work, embroidery and the process of creating art via wood burning'.[43] The resurgence of knitting continued to gather momentum as part of an international trend. Entries were down in 2017 but the standard and wide variety of exhibits was maintained, as it was again the following year to remain one of the larger displays of its kind in South Africa.

Under the active chairmanship of Henrietta Whelan entries increased again in 2019 with new interests emerging such as card making, scrapbooking and digital photography. More children were becoming involved in photography while spinning and weaving, once so popular, was apparently staging a comeback.

According to Di Fitzsimons, who concluded her long association with the section in 2018, during the previous decade the increasing incidence of home schooling seemed to produce high-quality work in a variety of fields. This was attributed to personal one-on-one training although overall the number of junior entrants was declining. Transport costs were making it increasingly difficult for distant women's institutes to compete and raising sponsorships was always a challenge. Ten-day demonstrations of various crafts were also becoming increasingly difficult to attract due to the expenses involved.[44]

Prizewinners in the Crafts and Home Industries Section included some familiar names from previous years, including the 2012 overall winner Avril Musgrave and her runner-up Ruth Aldridge. She was overall champion the following year with Una Taylor runner-up. Musgrave and Aldridge finished in the same order in 2014 but there was a new overall champion and runner-up the following year in Ann Duckworth and Tina Mossmer.

Northdene (Durban) won the women's institute competition in 2012 with Cowies Hill/Pinetown the winners in 2013, Swartkop Valley in 2014 and Premier Women's (Pietermaritzburg) in 2015. In 2013 the institutes mounted a special honorary exhibit of tapestries depicting the early history of Natal in acknowledgement of their federation's 85th anniversary. The section as a whole was awarded a gold medallion and trophy for excellence.

Laddsworth and Newton High secured the junior and senior school trophies in 2012, while Westridge High and Wartburg Kirchdorf were the senior and junior school champions in 2013. The former won again in 2014 with Maritzburg Christian School the junior champions. These two were succeeded by Newton Prevocational School and Maritzburg Preparatory in 2015. Claire Tucker from Ladysmith won the individual junior championship in 2013, 2014 and 2015.

By 2014 it was, however, noticeable that while the participation of veterans in the Show was increasing that in the junior and school categories was in decline as home crafts faded in the face of other demands on pupils' time. New senior and junior champions emerged in 2016 in Jane Henstock and Grace van der Velde while Cowies Hill again won the women's institute championship and Newton Prevocational and St Benedict the senior and junior school titles. Newton won again the following year with Howick Prep the junior school champions.

Nicola-Paige Hanegraaf was the individual junior champion in 2017 and 2018, Tina Mossmer the 2017 senior champion and previous winner Swartkop Valley the women's institute champions. Cowies Hill won that title in 2018, Westridge (Durban) the school championship and, at 85 years of age, Ruth Aldridge was the overall senior champion for a remarkable 17th time.

The **Commerce and Industries Section** also continued to thrive despite the prevailing economic climate. It comprised 420 exhibits in 2012 with a minimum of so-called flea market displays. The SAP, Safe City, and the Departments of Correctional Services and Justice, among others, combined to produce a Hall of Safety and Security which won the President's Trophy for special endeavour. Other specifically dedicated venues included the Daily News Special Occasions Hall, the Mercury Food and Festival of Fine Living Hall and the Witness Beautiful Homes Hall.

Several motor vehicle manufacturers were also present with good-quality exhibits although it was felt that those who crowded their stands with too many cars needed display guidance. The Eastern Bazaar south of the Main Arena lacked the quality of previous years when a wide range of worthwhile products from the Indian subcontinent were available at competitive prices but were now crowded by ordinary street-corner items.[45]

Most exhibitors in all sections of the Show who were canvassed continued to regard the outcome of their participation as either excellent or worthwhile but in 2013 the PCB withdrew its dedicated exhibit in Hall 1 because of a lack of commitment on the part of its sponsor the Department of Economic Development.[46] Even so, there were 450 exhibitors with a noticeable increase in the construction and materials handling categories in which distributors regarded the Royal Show as a very worthwhile marketing opportunity.

Among the dedicated displays the *Daily News* focused that year on an Absolutely Fabulous range of health and beauty products while the *Witness* presented a Hall of Technology. The Hall of Safety and Security was considered to be one of the most interactive and informative displays ever presented at the Royal Show. The KZN Sharks Board exhibit, which was part of the 20% new factor injected into the Show that year, was also highly informative and included shark dissections.

In 2014 Govan Mani, the last of the white goods exhibitors, relinquished its stand but there were still 437 exhibitors, the vast majority of those canvassed being satisfied that their expectations of the Show had been met. The 2 200m² Mercury Food and Festival of Fine Living Hall was by now well-established and confidently described as the country's 'largest dedicated exhibit of any kind', attracting an annual 105 000 visitors and 110 000 in 2015.[47]

There was also a hall focusing on the services and products of the province's rural entrepreneurs and a new display of interior decorations and furnishings. The bric-a-brac stands in Hall 3 were removed in 2015 to make way for an informative dinosaur display intended primarily for youngsters. Despite the additional R45 charge it attracted 14 000 visitors.

The need for a flea market dimension was still recognised by providing ample space for it in the Royal Market and Sharks Alley. Hall 1 was converted into an African market involving nearly 300 rural businesses with a spaza shop and 25 other stalls offering handmade crafts and garments as well as fresh produce. Another hall was dedicated to home décor while there was the usual range of exhibits from trucks and motor cars to fashion accessories and clothing.

After many years of participation, the *Witness* relinquished its stand but returned to its exhibition facility adjacent to the Main Arena in 2016–2017 with a security of tenure agreement. Its display then focused on its own 170th anniversary as the country's oldest surviving newspaper. It also returned to the Garden Show.

That year there were 404 commercial, industrial and service sector exhibits compared with 328 the previous year. It raised the prospect of having to reduce the size of stands if this number was to be increased. Hall 3 was dedicated to a successful Made in India Shopping Festival and Hall 10 to artisan craft exhibits from all over the province under the direction of the well-known local NGO Tembaletu.

The Mercury, Daily News and Safety and Security halls were as popular as ever with the demonstration kitchen attached to the first of these featuring the well-known Escoffier Jenny Morris. The redesigned Royal Food Court, previously Scott's Caterers, also attracted attention. While some exhibitors dropped out, Jeep and Mercedes Benz returned after an absence of several years and the Powerstar Truck range was a new addition.

In 2016 and 2017 there was still a strong presence of heavy-duty machinery but a reduction in construction and materials handling exhibits due to the economic recession, including the temporary withdrawal of Komatsu, which had considered the 2015 Show its most successful South African exhibit in recent years. With the non-participation of the Department of Correctional Services, the Made in India Shopping Festival was extended and a Pops, Moms and Tots Expo introduced.

The following year commercial exhibits were down again, from 432 to 428. However, some, most notably the Mercury Food and Festival of Fine Living at the heart of the Showgrounds in the Olympia Hall, continued to attract interest and the quality of exhibitors had significantly improved. The Daily News Hall gave hair products the limelight, including a programme of hair styling demonstrations, the Royal Indian Pavilion predictably displayed products from the Indian subcontinent and the Out of Africa exhibit focused on fashion wear from as far away as Cameroon and Nigeria.

An entire hall was allocated to Multichoice to promote its offerings, another to Ford for vehicle displays and yet another, in addition to the adjacent site, to the SANDF and SAPS for a first-time joint exhibit. Industrial and construction equipment manufacturers were again better represented with Caterpillar and JCB among others again well to the fore. There was a 34% response that year to the electronic questionnaire sent to exhibitors with 82% of them expressing a positive or very positive response that their expectations of the Show had been met.

Competition for the numerous sponsored trophies in this section was as keen as ever. The SAPS was a co-winner of the Royal Commercial Floating Trophy for the best and most outstanding display in an exhibition hall in 2013 and won it outright in 2014 and 2016 as well as the RAS/MUHL Floating Trophy for the best government department display in 2013, 2017 and 2018.

UKZN won the RAS Trophy for special endeavour in 2012 and the Royal Industrial Floating Trophy for the best and most outstanding display in an individual building in 2013, 2017 and 2018 while the Sheep and Wool Expo won it in 2015 and 2016. Weston Agricultural College won the President's Trophy for special endeavour in 2013 and 2017 as well as the Hulamin Floating Trophy for the best and most outstanding display on an open site in 2012 and 2018. So, too, did the SANDF in 2013 and 2014 and the RAS Trophy for the best government department display in 2012.

Toyota SA Motors won the WesBank Trophy for the most outstanding motor industry display in 2013, 2015, 2016 and 2018 as well as the Pietermaritzburg Chamber of Commerce and Industries Floating Trophy for the best South African industrial display in 2012 and 2014. In 2013 the Tea Merchant won the Mercury Trophy for the best display in the Mercury Festival of Fine Living Hall and in 2014 the PCB Trophy for a small business exhibitor.

Midlands Mascor won the Pietermaritzburg Chamber of Commerce Agriquip Floating Trophy for the best display of heavy-duty agricultural equipment every year between 2012 and 2017. Husqvarna South Africa won the David Wing Trophy for the best display of light agricultural equipment in 2012 and 2015, as did PMB Power Products in 2013, 2014, 2016 and 2018. Bell Equipment Sales SA Ltd won the RAS Materials Handling and Construction Equipment Trophy in 2012, 2013 and 2014.

The *Sunday Tribune* **Garden and Leisure Show**, as it was again now known, also had its fair share of winners and was still the most prominent of all the other activities held each year at the Showgrounds outside the Royal Show itself.

In 2012 the show ran for four days to take advantage of a public holiday with the improved attendance of 20 152 unfortunately restrained by the Saturday afternoon cloudburst mentioned earlier. The number of municipalities participating continued to grow with uMhlathuze (Richards Bay) and Khara Hais (Upington) joining their ranks. The latter was judged best newcomer among the 76 designer gardens on show ranging from 1m² to 160m² in size. By then the event was well established as the premier annual showpiece for South Africa's municipal parks departments.

The city of Cape Town was again the People's Choice but Arthur Mennigke's panel of judges, applying their usual international standards, awarded the Cramond Garden Club a score of 98% and gold with laurel as the best on show even though it was only a 2m² exhibit. This caused consternation among landscapers who contended that macro and micro displays should be judged separately.

Cape Town's exhibit maintained its prominence by being the People's Choice again in 2013 and 2014, while eThekwini (Durban Parks) and uMhlathuze were also acknowledged for innovation and improvement. Feature gardens which regularly attracted the judges' attention included the Gordon Stuart Landscaping/*Sunday Tribune* exhibit which in 2013 exceeded a score of 95%.

The following year the judges, as always quite independent of the RAS, abandoned their usual percentage score ratings and awarded gold (seven) and silver (seven) to whichever displays they considered worthy. Award-winning garden club exhibits included Just Boutique and Pot Pourri.

Youth participation was again encouraged with grades 4 to 12 school groups, some from rural areas, competing as before in their own mini-exhibit category and Northbury Park Secondary School winning gold medallions in 2012 and 2013. A feature of the Schools Hall was the Bug Hotel, a cross-sectional display illustrating the various habitats of different insect species.

Tanya Visser again compéred the entertaining 'ready steady plant' competition in which four teams demonstrated their ingenuity by constructing 2x4m gardens in 20 minutes using identical but surprise material. The Voortrekker School boys also contributed to the show each year by providing a welcome wheelbarrow service to assist visitors in transporting their purchases to the car parks. Some exhausted customers were tempted to sit in the barrows themselves!

Every year there were additional attractions such as a new focus on the environment relating to issues of conservation and sustainability in the 900m² Happy Earth Hall. The 1 000 Hills Chef School conducted classes in the demonstration kitchen on the use of fresh fruit, vegetables and herbs, while the Orchid, Floriate (flower arranging) and Quilters halls mounted colourful exhibitions. So, too, did the Ilovo Sugar-sponsored floral cake icing competition, which in 2012 attracted 400 entries and featured the reproduction of a 2m Kapok tree in full bloom made entirely of sugar.

That year Skal International, the association of travel and tourism professionals, hosted an exhibition showcasing KwaZulu-Natal's tourist attractions with the Karkloof Farmers' Market a welcome addition. The annual displays of leisure activities spread over two arenas included archery, caravans, motor vehicles, a 4x4 course and dog agility championships. There were also art exhibitions, apiarian exhibits, a Kidz Zone for children and in 2014 an international clivia conference as well as a reconstruction of South Africa's award-winning exhibit at the recent Chelsea Flower Show.

In these ways the Garden Show maintained public interest with attendance improving in 2014 from 16 073 to 17 343. It was also due partly to almost R3 million worth of extensive advertising. This included 2 500 pamphlet handouts and the services of an internet marketing company to develop an online media campaign involving 45 000 bulk SMSs.

Unfortunately, for all its popular appeal by 2013, its 38th year, the Garden Show continued to be a loss leader with a R92 272 deficit.[48] Keynote exhibits, especially those from other provinces, were expensive and required subsidies and/or grants which the RAS could not go on providing. When the show reflected a substantial loss for a third consecutive year, in this case R464 192, it seemed unlikely that the Society would continue to host the event and the partnership with the *Sunday Tribune*, which had held the naming rights for a decade, ended.

After discussions with all interest groups Lonehill Trading, publishers of *The Gardener* and *Die Tuinier*, undertook to underwrite future losses and Tanya Visser assumed the challenging role of co-owner and show director for the next five years. The RAS became, in effect, 20% partners by providing the venue, facilities and logistics while 80% of any profits were to accrue to Lonehill Trading.[49]

Tanya Visser began her life's interest by assisting her father in his vegetable patch and helping her parents around the garden. She studied horticulture at Natal Technikon and acquired further practical knowledge working in a garden centre.

After several years in the industry she moved into publishing, producing gardening books and becoming the owner/editor of *The Gardener* and *Die Tuinier*. She also became a nationally known TV personality presenting her own show 'The Gardener'. As a well-established garden designer, horticulturist and motivational speaker she was elected president of the South African Nurserymen's Association and joined the board of Food and Trees for Africa. She continued to grow her own vegetables.[50]

Tanya Visser applied a very business-like approach involving more commercial nursery industry participation and careful control to generate a moderate but significant turn-around. In 2015 a reinvented Garden Show produced a R54 277 surplus after Lonehill Trading's profit share had been deducted. It was the first surplus in five years, and the *Witness* re-established its naming rights to the event as well as holding its 170th banquet at the Showgrounds.

With 180 exhibitors occupying over 12 000m² of open space and 4 000m² under cover the Show's financial recovery was greatly assisted by an ex-gratia R150 000 grant from the Msunduzi Municipality. This was a timely recognition of the value which it had added to the city over many years. It was estimated that 43% of visitors to the show were from greater Durban, 35% from Pietermaritzburg and the Midlands, 20% from elsewhere in South Africa and 2% from overseas. Approximately 50% of them were between 18 and 40 years of age, 42% were older and 8% were younger.[51]

The slight decline in attendance from 17 343 in 2014 to 17 160 was probably attributable to the weather during what was the hottest Garden Show ever recorded. This did not discourage the RAS from proclaiming that it was now 'the largest horticultural event of its kind in the southern hemisphere'.[52]

While Effekto, Protek, the RAS and *The Gardener* sponsored new trophies as part of the show's revamped image municipal parks departments continued to anchor it, including six new designers in 2015. The positioning of some exhibits was altered to improve pedestrian traffic flow and, for the first time, a system of peer judging was introduced. This involved previous gold medallion awardees overseen by David Davidson, designer and multiple winner with the South African exhibit at the Chelsea Flower Show.

Perennial top achievers Cape Town won gold, as did Gordon Stuart Landscaping and the Midlands Rose Society. The RAS and Best on Show eThekwini Municipality won gold with laurels and did so again in 2016. The latter eThekwini entry, which Francois Lenferna designed, was only the second recipient in South Africa of an unqualified 100% score. It won the awards for Best on Show Feature Garden, the People's Choice and Best Municipal Feature Garden, securing the last title again in 2017 and 2018 with Mbombela (Nelspruit) the People's Choice in 2017.

That year Gordon Stuart Landscaping/Fresh-line Flora and Parklane Superspar were adjudged the best feature gardens winning gold with laurels while Dave Moore won gold for the RAS exhibit. In 2018 he secured the Effekto Care Floating Trophy for the best feature garden with a stunning design using orchids, ferns, grey beard, pink and white blooms and fountains.

In 2018 eThekwini was again adjudged the Best Municipal Feature Garden and Msunduzi won the CEO's award but the display of the Midlands Rose Society was the most popular among the general public. Among the garden clubs Pot Pourri won gold with laurels in 2015 and 2016, with Hilton and Cramond winning silver, both in 2016. The first was the best garden club feature garden in 2017 and Pot Pourri tops with gold in 2018 and 2019. That year for the first time five amateur enthusiasts were chosen from several entrants to build Blackwood's-sponsored 9m² theme gardens with a given selection of materials.

The Garden Show's focus on youth was maintained along with the emphasis on sustainability and the green industry. In 2015 eighteen schools were allocated 1.5m² plots in the Happy Earth Hall and were expected to include a recycling dimension in their designs using products which ABI supplied. Chislehurst, Allandale Primary and Gert Maritz Primary all won gold with the last winning the Coca-Cola-sponsored schools design competition in 2016 and 2017 and Clarendon the recycling project in 2016. Hilton Montessori won it in

2018 when HS Ebrahim School produced the best schools mini garden.

As before, there were numerous other attractions each year. The Plantae Orchid Club, the Bonsai and (for the first time) Cycad Society all presented exhibits in 2015. The Go Green Hall, also included for the first time, was more commercially focused with exhibits relating to pre-paid meters, solar energy and hybrid motor vehicles. A Food Hall was introduced in the members dining hall and terrace.

Familiar features included Floriate with a range of different sized exhibits which won gold with laurels, followed by the Protek Floating Trophy for the best specialist display in 2018. There was also the Baking and Creating with Illovo exhibition of icing while an animal farm and mini planetarium were included in the Kidz Zone.

In 2016 there was a national Floriate Indaba and Exhibition involving 200 delegates, a fashion parade with a horticultural theme in the Parks Hall, a Mums and Tots display of casual wear and a World of Dogs and Cats exhibition to complement the usual dog agility and canine judging competitions. A dedicated Heritage Day braai area was demarcated in partnership with Bluff Meats and the KZN Bee Farmers Association won the trophy for best specialist display with an exhibit that explained the life cycle of bees, so vital to horticulture.

The Crafts Hall was opened for the first time outside a Royal Show period to display crafts, glassware, jewellery and botanical art works. Another Garden Show first was the occupation of the entire 10 000m² Main Arena for a full programme of outdoor events. Subsequent new attractions included an interactive Working with Wood Expo for DIY enthusiasts, a Hall of Exotic Plants including bonsai, clivias, orchids and water plants, a floral art fashion show featuring the KZN Floral Union team and a school recycling project fashion show.

The 18 002 attendance in 2016 was an improvement of 842 on 2015 (a four not a three-day event) despite initially inclement weather. In addition to other marketing promotions, local public interest may have been boosted by the 32-page guide that the *Witness* generously published. It was considered the best supplement ever produced in support of any event held on the Society's premises. Pleasingly, the show generated another surplus, R262 450 overall.

It only attracted 18 125 attendees in 2017 although it was a four-day event intended to take advantage of the Heritage Day public holiday. This was not favoured by temperate weather but nevertheless produced a R391 759 surplus. There were 156 exhibitors, some of whom reported best-ever sales, and nineteen feature gardens, but there was a decline in municipal participants due to financial constraints.

In January 2018 the partnership between the RAS and the publishers of *The Gardener/Die Tuinier* was extended with R15 000 payable to the former to cover administrative costs. Unfortunately, attendance declined to 15 595 that year due to bad weather with 30 mm of rain reducing the show to a two-day event. The anticipated surplus of R150 000 was also substantially down on 2017, but still generated a commission of R30 000 for the RAS.

Several of the 170 exhibitors reported record sales and despite the decline in municipal representation there was an increase in the number of private and corporate landscapers as well as the usual variety of other attractions. These included the South African Floral Union Conference and the inaugural All Africa Cup for Floral Art competition, as well as the country's premier cake icing competition involving more than 130 exhibits, a Clivia Society display, the now usual Schools Gardens and Happy Earth exhibit with the emphasis on the prevention of environmental degradation, artwork and quilts and a tool expo together with the Working with Wood exhibition.

Favourable spring weather greeted the 44th Witness Garden Show in 2019 which attracted more than 200 exhibitors and 18 264 visitors with good sales being recorded despite the prevailing economic recession. Not for the first time, the judges declared several of the feature gardens to be on a par with those of the Chelsea Flower Show. Two new competitions were included in this section: the Great Big Show Garden Design offering a R100 000 prize for professionals and the Blackwood's Backyard Show Garden Design for everyday gardeners with R10 000 worth of vouchers as first prize.

Shaun van Huysteen won the latter while Dave Moore won the former as well as gold with laurels, the Efekto Care Floating Trophy for the Best Feature Garden on show and the People's Choice award. The Bonsai Society won the Protek trophy for the Best Specialist Display, the Pot Pourri Club was the only contestant for Best Garden Club Feature Garden, and eThekwini was again voted the Best Municipal Feature Garden.

As before, Coca-Cola sponsored the school section with Shri Vishnu Primary producing the best recycling project and Allandale Primary winning the overall competition. For the 15th year in succession pupils from neighbouring Voortrekker High School assisted visitors to transport their purchases to the car parks by serving as barrow boys.

That year the Clivia Club celebrated its 25th anniversary while several gardening experts including Tanya Visser participated in a series of free sponsored talks and demonstrations. The Working with Wood Hall and

Drakewoods Tool Expo proved to be of great interest, as was the Spring Dog Show and the Border Collie Rescue performance. The popular Starke Ayres Kidz Zone featured a 4Kids Carnival with numerous activities and free planting demonstrations.

By 2019 the new reinvented Garden Show could justifiably claim to be 'bigger and better' than ever before. It still attracted the majority of its visitors from the greater Durban area as well as others from as far afield as Johannesburg and Cape Town with only a minority now drawn from Pietermaritzburg and the Natal Midlands.[53]

Other activities held at the Showgrounds, apart from the Garden Show, were gaining increasing importance in the Society's calendar and in its financial returns. These included what had become regular annual events such as the official opening of the provincial parliamentary session, the UKZN graduation ceremony and the Department of Health nurses' graduation.

There were also occasional functions such as the KwaZulu-Natal provincial COSATU conference in 2012 and the Premier's Freedom Day concert, the SASSA registration process, the Oxford University Press conference, the Mercedes Benz Truck Division Roadshow and the provincial launch of the new range of John Deere tractors, all in 2013. There were the usual regular bird, poultry and rabbit shows. Some of these were quite large, such as that of the Oxford Gamefowl Club which the Natal and Coast Poultry Club hosted in 2014 and again in 2015, 2016 and 2018.

The annual Feather and Fur Day held in August 2020 attracted support from all sections with the Poultry Club being joined by the KZN Fancy Pigeon Club, Natal Rabbit Club, Pietermaritzburg Budgerigar Club and Pietermaritzburg Canary and Cage Bird Club. The Rabbit Club also held a pet show for the first time, including bunny hopping demonstrations, as well as its second championship of the year. The Showgrounds-based KwaZulu-Natal Bee Farmers Association hosted the Royal Honey Festival in August 2019 which, as before, focused on promoting public awareness about the ecological importance of bees by way of lectures and practical demonstrations.

There were numerous equestrian events. In March 2012, for example, a three-day fundraising Royal Horse Festival was held at the Showgrounds with the proceeds going towards hosting the South African Pony Rider championships in July at the Shongweni Club. There were many first-time junior entrants competing in the presence of parents and grandparents who themselves had previously participated. There was a four-day Horse Festival in 2013 in which all four disciplines were well represented. In 2019 the 58th annual Pietermaritzburg Pony Show involved 46 horses and continued to be regarded as a springboard for many of the country's future top riders.

In 2012 the RAS hosted 48 government and private sector representatives at a marketing presentation, Showgrounds tour and lunch to promote greater awareness of its facilities.[54] That year there were 92 conferences, functions and other events held on its premises involving 170 000 visitors. This made it one of the top five busiest venues in South Africa and contributed more than 26% of the Society's income, providing ample incentive for ongoing marketing and upgrading efforts.

There was a further increase in functions held in 2013 when revenue increased by 19% to exceed R4 million for the first time. The RAS was by then accommodating about five functions a week and catering for 185 000 visitors a year. In 2014 it also successfully hosted the South African Sugar Technology Association conference. Usually held in Durban, it was 'the first international conference secured in open competition with blue chip South African venues'.

An association representative subsequently described the venue and the catering as 'excellent' but observed that 'just too much food' had been provided at mealtimes! Further, that security could be improved as two 'hoboes' off the street had joined the lunch queue before being asked to leave. Subsequent investigation established that the hoboes were one of the Society's own 'highly esteemed' contractors and his son who had been attracted by the generous spread provided! More seriously, the conference did expose the need for the Showground's WiFi system to be upgraded to cater for 'mass usage'; that is, more than 60 users at any one time.[55]

In 2014 the city gave the RAS an award of excellence for its conferencing facilities. There were more big events in 2015, including the State of the City Address, a three-day Royal Horse and Hound Festival and an international grasslands conference. By then there was an overall decline in use of the Showgrounds and a 12% contraction in revenue, in common with conferencing facilities elsewhere.

This was due to the prevailing economic recession, provincial government's decision to curtail expenses, the cancellation of several functions following the death of ex-President Mandela and the subsequent distraction of the May general election resulting in several other governmental events being postponed or cancelled.

Attracting conferences and other events continued to rely as much upon favourable word-of-mouth reports as on the media, whether social or otherwise. Perhaps somewhat belatedly, in 2016 the Society invested in a data projector, laptop and screen. Business improved again that year with several of the regular major events returning to the Showgrounds along with new functions. There were 800 in all, attracting 160 000 visitors.[56]

These were as diverse as a Mascor tractor launch, an ANC conference, an Msunduzi service awards ceremony, a medical aid employee awareness campaign and the annual Pony Show in the Main Arena with more than 60 entries. As before, there were numerous private occasions drawn to the Showgrounds by considerations of convenience and safety. Their 1 kilometre proximity from central Pietermaritzburg, 2 kilometre distance from the N3 motorway and easy 45 minute access from greater Durban were all added attractions, together with adequate parking.

The Showgrounds also offered a variety of venues. It could accommodate up to 25 000 visitors in the outdoor Main Arena and 1 500 in the cattle arena while its indoor options provided for between 30 and 3 000 persons seated. All venues could be furnished and decorated as required with full catering facilities 'from a spit-braai to silver service dining'.[57]

At the Showgrounds epicentre the Olympia Hall was aesthetically revamped to assume a cathedral-like appearance to accommodate 3 000 international congregants, including Vatican representatives, who celebrated the 200th anniversary of the Roman Catholic Order of Mary Immaculate (OMI). Some 4 000 mourners attended the Main Arena funeral service of Catholic Bishop Barry Wood.

Among numerous other major events there were 3 000 attendees at the three-day official opening of the KwaZulu-Natal legislature in the presence of King Goodwill Zwelithini. In 2017 the RAS provided the venue for the International Mounted Games and for the Starlight Jazz Festival. Several firms, especially in the motor industry, still used the Society's facilities for logistical support and storage.

In 2018–2019 there was another round of contraction as parastatal and provincial government clients in particular again cut back on non-essential expenditure. Not all the big events were cancelled. The Bridal Fair, the National Honey Conference and the Honey Festival exceeded expectations. The last attracted 1 500 visitors and the Starlight Jazz Festival more than 9 000. The nurses and UKZN graduation ceremonies were still held as was the three-day opening of the provincial legislature with its 3 000 delegates.

The first two of these three events were hosted again in 2019 but the last of them, which was initially due to be held in the downtown parliamentary buildings as an economy measure, eventually took place in the Showgrounds in a substantially reduced form. Other major events included the Department of Agriculture's conference and dinner for 1 500 invitees and the Department of Education Long Service Awards, previously held at Durban's ICC.

Such events were as important as ever to the Society's financial well-being, but income from that source declined by 22% between 2018 and 2019 from R4 453 636 to R3 486 497.[58]

Finances were remarkably sound considering the prevailing recessionary economic conditions. Moreover, in 2012, for the first time in its 161-year history, the Society was required to pay income tax in terms of new SARS regulations published in 2008, even though no individual or organisation benefited directly from any surpluses that it generated. Fortunately, a R800 000 payment was followed by a R330 000 refund for the 2012 and 2013 tax years. The RAS also became VAT registered, being applicable to income such as gate fees and parking tickets.[59]

In addition to substantial increases in electricity charges there was an expensive misunderstanding with the municipality concerning the phasing out of its previous financial grant which came to an end in 2012. Yet another unexpected expense arose from a power surge that necessitated the replacement of the Society's switchboard, which was not covered by insurance.

Despite these changing circumstances and setbacks, in 2012 the RAS still managed to achieve an all-time record financial surplus. While income from the Royal and Garden shows was better than expected this was largely due to a significant increase in income derived from other Showgrounds functions. It also enabled the Society to more than meet its bond obligations in respect of the 2006 land swap.

As a consequence of losing its tax-exempt status the RAS began to look for appropriate tenants who might permanently occupy some of the facilities within the Showgrounds. In 2012 the first of these, one of its sponsors ABI, signed a medium-term lease to establish its regional marketing office and in 2014 entered into a second lease for more office space in Sharks Alley.

Kentucky Fried Chicken opened an outlet at the Showgrounds and assumed branding rights to what became KFC Tower, previously the Vodacom Tower. The cost of subsequently refurbishing the tower was shared between KFC, Vodacom and the RAS. Ola Ice Cream replaced Nestle Dairy Maid in terms of an agreement covering exclusive rights in the Showgrounds over the next three years for a R56 000 fee.

Surprisingly in the prevailing economic climate, in 2013 the Society produced a before tax surplus of R3 220 952, which constituted a 53%

improvement on the previous year. Although the Royal Show was the primary source of income, other events outside Show time again made a significant contribution. This was attributed not so much to marketing as to word-of-mouth recommendations by satisfied clients. That year the longstanding grant the city had provided for electrical consumption was terminated.

The financial year ending June 2014 was beset by a decline in income from the Royal and Garden shows and from conferences and functions. Pleasingly, the Society still managed a 6.5% increase in pre-tax earnings and a respectable bottom line thanks to an income distribution from the Royal Mews Trust and tight cost controls.

That year the R6 million bond raised to purchase the Showgrounds outright was paid off, seven years in advance of its fifteen-year loan term. It was a welcome release considering the negative macro-economic trends that now beset the RAS in common with similar societies elsewhere.

These included an ongoing decline of the business confidence index and depreciating currency with its adverse effects on costs, the persistent uncertainty concerning government's intentions about land reform, the demands of trade unions and the likelihood of further drought conditions following the strongest El Niño phase in 50 years. In addition, the demographic pattern of the region was changing dramatically, the number of volunteer assistants, the traditional backbone of the RAS, was declining, and Showground infrastructure was aging.

Moreover, as president Mike Moncur recognised, 'the relevance of agricultural shows worldwide' was 'increasingly being questioned' in terms of cost and efficiency. One-on-one, face-to-face interaction aside, it was being argued that there was more to be gained 'via the web and the social media than one could achieve at a show'. In response, the Society formed a sub-committee to investigate the possibility that even with more vigorous marketing its current modus operandi might eventually have to change.

A SWOT analysis was undertaken to examine its present condition and consider possible future options. Among its recognised strengths were its freehold ownership of the Showgrounds, a capacity to generate income as demonstrated by the Royal Mews Trust, the loyalty and expertise of its management and volunteer helpers, its proud history and reputation as well as its ability to attract sponsorships.

Its weaknesses included the necessity to pay rates without any compensatory municipal rebate, the disappointingly negligible support received from the provincial Department of Agriculture, its aging infrastructure and high maintenance costs, declining volunteer numbers and relatively static Show attendances, as well as the fact that the centrally important Royal Show livestock sections were not revenue generators.

The opportunities identified included the Society's potential to generate additional income by hosting more conferences, business functions and dedicated shows; the possibility of entering into strategic partnerships, such as that recently negotiated in connection with the Garden Show, in order to finance some of its activities; and the renting out of certain of its facilities to permanent tenants, as was the case with the Land Bank. There was also the prospect of allocating more of its property to commercial development, as had already taken place along Chatterton Road, or of even selling the Showgrounds in their entirety, investing the proceeds and using them to support appropriately selected projects.

Among the immediate threats which it was believed the RAS now faced were the declining number of farmers and increasing on-farm shows, expanding on-line business, new technology such as artificial insemination which reduced the need to exhibit bulls and the increasing costs incurred by exhibitors in participating at the Royal Show. Other broader concerns were changing demographic patterns and values, fluctuations in the economy and weather, rising crime levels and the possibility that if the Showgrounds were rezoned as a commercial area it would result in a rates burden which the Society would be unable to sustain.[60]

Meanwhile, in 2015 the Royal Show contributed 62% of the Society's gross income from stand rentals, gate takings and sponsorships. Conferences, functions and other events contributed 23%, the Garden Show 4%, financial investments and administration fees 7% via the Royal Mews Trust, other sources 3% and membership 1%.[61]

Despite the unfavourable ongoing economic conditions, the RAS enjoyed a pre-tax surplus of R1 529 000 in 2016, amounting to a 26.7% increase on the previous year. This was attributable not only to cautious expenditure but also to a successful Royal Show and improvement in the finances of the Garden Show. Any thought of shortening the length of the former was dispelled when Terry Strachan pointed out that weekend gate takings equated to 70% of the R4 040 700 paid at the entrances.[62]

The financial contribution of the Royal Show was nevertheless disappointing the following year. This partly accounted for a decline in the pre-tax surplus to R1 413 855 even though conferences and functions experienced an 8.3% growth. The financial contribution of the Royal Mews Trust was also vitally important, but there was concern that this cross-subsidisation might not be sustainable in the long-term. Fortunately, the outstanding bond which had been raised with FNB was fully repaid by October 2017, two years ahead of due date.

In September 2017 the executive re-visited the SWOT analysis undertaken the year before and acknowledged that it was 'unlikely that the business of the Royal Agricultural Society, in its own right, will return to generating meaningful surpluses'. It would therefore be amenable 'to receiving and considering any additional developmental proposal' provided its 'constitutional commitment to being a conduit of support for agricultural endeavour is not compromised'. The executive stressed that 'whatever transpires going forward the Society is honour, ethically and legally bound to never relinquish its primary role to support agriculture'.

Manco continued to re-appraise its current business model and in January 2018 it informally discussed possible future scenarios with Peter Miller, a longstanding member and ex-MEC for Finance. Later that year it affirmed its conviction that the Royal Show was 'a prestigious and acknowledged brand' and that the survival of the RAS was 'non-negotiable'.

It speculated that its business model might eventually mirror those of the Royal Highland and Royal New Zealand shows in which 'large tracts of lawn, to the exclusion of built infrastructure, could form the basis of future exhibitions'. To that end the acquisition of land somewhere in the Midlands close to Pietermaritzburg might provide a venue for agricultural exhibitions as well as events similar to those currently held at Nampo. There was concern about declining membership and the difficulty in securing individuals who were willing to serve on the Society's various overseeing committees.[63]

Meanwhile tight control was maintained over expenditure. This increased by 4.8% in 2017 and would have been as low as 2.9% but for the necessity of reinforcing the Dorpspruit riverbank with gabions. The increase in expenditure was reduced to a mere 0.6% by 2018 (compared to a CPIX of 4.6%) and the RAS achieved a creditable pre-tax surplus of R1 203 320. This was despite a marked decline in income from conferences and other functions in trading conditions which CEO Terry Strachan described as 'the most challenging in 21 years'.[64]

A R200 000 overcharge for water consumption was eventually reversed and that year Clover moved its Pietermaritzburg offices to its premises in the Showgrounds, joining the sales office of future sponsor Coca-Cola in Block C. Mitchells Plumbing entered a medium-term lease agreement for the ex-

Govan Mani building while Ola Ice-Cream and CK Cigarettes confirmed their exclusive rights in the Showgrounds. They were subsequently followed by Mini Do-nuts/Chip Twister and Squirrels Bavarian Nuts. That year the RAS was registered as an official service provider to the eThekwini Municipality which subsequently assumed ownership of Hall 4 to showcase its facilities.[65]

By 2018 the largest item of expenditure was still salaries and wages, amounting to 25% compared to 24% in 2012, followed by other (13% compared to 21%) and electricity, water and rates (11%, the same as in 2012). Happily, that year the Society had a R1 million credit with SARS due to overpayments. The loss in 2018–2019 of Standard Bank (after 50 years), Coca-Cola Sales and Marketing and Hindvani FM as tenants was accompanied by long-term rental renewals by Clover SA, Mace Safety Solutions and the PCB, which had occupied Chamber House since 1984.[66]

The major sources of the Society's income in 2018 were show rentals (36% compared to 27% in 2012), gate takings and parking (29% compared to 33%) and grounds usage (21% compared to 26%). The last two sets of declining percentages pointed to the prevailing recessionary conditions, as did the reduction in income from sponsorships and donations (7% to 4%).

The financial returns for 2019 reflected a refund of R350 000 in transfer duties relating to the Bank Park subdivision but also a marked decline in surplus earnings from R1 203 320 in 2018 to R581 346. This was particularly adversely affected by the 22% decline in income from conferences and other events outside of Show time. This made it even clearer that, but for the income generated by the Royal Mews Trust and in the absence of any economic upswing, the deficits being incurred by the RAS could not be sustained indefinitely and that the adoption of some new business model might become essential.[67]

Sponsorships, although currently in decline, were nevertheless as important as ever to the Society in terms not only of its financial well-being but also its public image. In 2012 and again in 2013 it was grateful to acknowledge ABI, Telkom and the Land Bank as major supporters of the Royal Show. In 2012 the provincial Department of Agriculture unexpectedly withdrew its corporate sponsorship of that event. This made it the only exhibition of its kind falling within the ambit of the RASC that did not have the support of the relevant government department in its part of the world.

Fortunately, as the RAS readily acknowledged, in 2014 Standard Bank 'continued to contribute significantly to the Royal Show by way of being the premier sponsor of the livestock Supremes'[68] and the *Sunday Tribune* maintained its commitment to the Garden Show. That year Telkom withdrew its corporate sponsorship of the Royal Show due to the recessionary economic climate. The Land Bank remained and in 2015 ABI renewed its corporate sponsorship of the Show for the next three years in the amount of R695 000 while the *Witness* returned as the major sponsor of the Garden Show.

Among the now numerous formal and informal social functions that took place during each Royal Show the most prestigious was the Standard Bank Gold Cup Dinner that usually followed but sometimes preceded the cattle parade and presentation of trophies. In 2016 the Land Bank became the major sponsor of this important annual event but in 2017 it withdrew its corporate sponsorship of the Royal Show. That year the Gold Cup Dinner and the traditional corporate sponsors evening were combined with 300 guests in attendance. There were 200 at the farmers' barbeque in the cattle arena which replaced the sheep supper and was co-sponsored by Sanlam and *Farmer's Weekly*.

East Coast Radio and First National Bank joined Coca-Cola as corporate sponsors of the 2017, 2018 and 2019 Royal Shows. After an association of 167 years the RAS claimed to be possibly the bank's oldest South African client. In addition to these three generous supporters Standard Bank, the *Witness*, the *Mercury* and RSG were also acknowledged for contributing more than R50 000 in cash or kind.

Standard and the RPO continued to contribute substantially to the Show's livestock sections but, after a presence of more than half a century the former, as previously mentioned, gave notice that it would vacate its Showgrounds premises at the end of 2018. Even so, in 2019 the RAS still boasted 48 corporate and other organisational sponsors in addition to numerous individual donors.[69]

The **Royal Mews Trust**, even more so than sponsorships, was still vital to the Society's well-being. It was now well-established as the 'property developmental arm' of the organisation and an 'independent cost centre' in its own right. In 2012 it enjoyed its best financial year ever when it recorded a surplus of R1 218 564 before taxation that constituted a 66.5% improvement on the previous year.

As before, there were ongoing discussions with municipal and provincial officials to explore the possibility of developing part of the Showgrounds as a conference, if not a hotel, site. The former indicated support in principle for the proposal which would have involved a land swap but, after preliminary talk of a R100 million grant from the provincial government, the MEC for Finance Ina Cronje indicated that it was no longer under consideration.[70]

Meanwhile, there was concern about one of the Trust's tenants, the Dros Restaurant, which began trading in November 2011. This was six months behind schedule and, in common with others in that industry, it was finding economic conditions challenging. After extensive negotiations, by 2013 it had unfortunately become necessary to terminate the restaurant's occupancy and have its operating entity liquidated in view of rental arrears.[71]

Despite the heavy costs incurred the Trust improved its pre-tax revenue that year by 24% over the previous year to R1 515 000. The following year Hereford Financial Services, part of the Liberty Group, committed to a long-term lease of the same premises and to investing R1.1 million in refurbishing them.

A much broader issue that required serious reconsideration was the continued existence of the Trust itself in view of the new SARS tax regulations. As the president Iona Stewart explained:

'The primary rationale behind the creation of the Trust was to compartmentalise operations under the Society's ambit whereby those involved in shows and exhibitions (and tax exempt at the time) were separated from those of a taxable commercial nature. To the extent that the Royal Agricultural Society of Natal has since been adjudged a taxable entity, the perpetuation of the trust now appears somewhat meaningless.'[72]

A Price Waterhouse Coopers investigation was duly launched and the three properties under the Trust's custodianship – Bank Park, Dros Restaurant/ Hereford Financial Services and McDonald's Garden Centre/Blackwoods Nursery – were valued at R20 million. This raised the possibility that the capital gains tax involved in transferring their ownership back to the RAS could be prohibitive and a final binding decision was sought from SARS.

The latter subsequently rejected the Society's request to terminate the Trust without paying substantial capital gains. It was therefore decided to maintain the status quo and to transfer the sub-divisions from the RAS to the Trust in terms of the sale agreement entered into some years previously, except (initially) for that occupied by Blackwoods due to the cost of separate reticulation. Apart from tax considerations this substantial expense had been a major reason for considering the termination of the Trust in the first place.[73]

While Bank Park awaited a rates clearance certificate from the municipality, the proposed transfer of the sub-divisions that Hereford Financial Services and Blackwoods Nursery occupied was more complicated. This was due to the public utilities they both shared with the Showgrounds which would be expensive to separate and duplicate, an estimated R2 million in the latter case.

For a variety of reasons, the proposed transfers could not be immediately effected, but Prospect SA/Nick Proome was granted the right to conduct a feasibility study and have first right of refusal with regard to a possible extension of Bank Park westwards, north of the flood control canal and south of Hyslop Road. In the interim an updated valuation was agreed for the ensuing five-year period to ensure that an appropriate rent was levied and the lease agreement in connection with the nursery was also redrawn.

In August 2017, with all outstanding issues settled, transfer of ownership of Bank Park from the RAS to the Trust was at last finalised in terms of the agreement of sale of August 2007. A substantial refund of R424 000 in transfer duties was subsequently received from SARS. By 2019 the transfer of the properties that Hereford Financial Services and Blackwoods Nursery occupied still had to await the completion of public utilities independent of those serving the Showgrounds.[74]

Relations with the three primary tenants continued to be cordial and, after a three-and-a-half- year hiatus the former tenant, the Dros Restaurant, undertook to pay its outstanding debt of R888 031 piecemeal. The Trust eventually accepted R40 000 'in full and final settlement' of the longstanding debt but in 2019, R27 000 was still outstanding.[75] The Royal Mews Trust nevertheless remained financially successful with a pre-tax improvement in profit of 43% in 2015 over the previous year and 15.7% in 2016, amounting that year to R1 780 990, R1 865 053 in 2017, R2 113 670 in 2018 and R2 329 600 in 2019.

Financial and other pre-occupations aside, the Society was always careful to maintain, as best it could, links with associations and external events that were relevant to its own avowed mission.

The **profile of the RAS** continued to be maintained in part by its representation at overseas and local conferences. Iona Stewart attended the Botswana Show on the Society's behalf while she and next-generation candidates Mark Stewart and Courtney Wood represented it at the 2012 biennial RASC conference in Zambia where she chaired a plenary session and Mark Stewart presented a paper.

As a cost-saving measure, following a 27% depreciation of the Rand in twelve months, for the first time in 21 years the RAS was not represented at the next meeting held in Brisbane or at those held in 2016 and 2018. The Society declined an invitation to hold the 2016 conference because it would not generate a revenue stream but the reciprocal benefits of membership nevertheless continued to be appreciated. This was reinforced by the visit of Michael Lambert, honorary secretary of the RASC, after which the RAS was granted a 10% discount on its 2014 subscription.

On his return in 2016 Lambert reduced its annual subscription by 25%, amounting to R9 800, after being informed that the weakening Rand was curtailing all expenditure involving foreign exchange. He explained that peer societies elsewhere in the world were experiencing similar challenges but that, unlike Pietermaritzburg's Royal Show, they were 'generally well supported by local and regional government'.[76]

Nearer to home in 2012 and again in 2013 at the annual SAMIC dinner in Vryburg the Society received the award for the 'largest mixed Agricultural Show in the country' [77] and, in the latter year, for achieving the highest auction prices in 2012–2013 for beef (R110 per kg) and lamb (R375 per kg). At the 2014 dinner the Royal Show was again acknowledged as the largest agricultural show in South Africa and for attaining the highest auction prices in 2012–2013; that is, R475 per kg for lamb and R110 per kg for beef, the latter still being a national record. It was also adjudged the best pig show in the country.

In 2015 the Royal Show was recognised yet again, this time as the top commercial pork show in 2014–2015, South Africa's largest agricultural show hosting a national carcass competition and for setting a new national record price of R150 per kg for a beef carcass. The Royal Show was less successful the following year, but did win the pork-show evaluation category with 93.61%.[78]

In 2017 it won more awards, among others for being the country's largest agricultural show hosting a national carcass competition, the third best lamb show, the best commercial pork show and the best emerging pork show as well as several individual commercial and emerging farmer pork show categories.

The following year and in 2019 the Royal Show was again recognised as being the largest agricultural show hosting a national carcass competition and the best pork competition in South Africa. It was also acknowledged for producing the champion commercial pork carcass (NHELC Farm), first and second reserve champions (both Craig McCord), the national group champion commercial pork carcass (Butt Farming) and first and second group champion (both Craig McCord). The ten champion awards won in 2019 was probably a record for any single show.

In 2018, as in previous years, Iona Stewart represented the Society at the annual Kwanalu Congress while Andrew Adams did so at the SAMIC Awards dinner. In 2019 the Royal Show became the first show to win all the commercial SAMIC awards for pork as well as a national champion award in the category for emerging farmers.[79]

The **administrative and grounds staff** did not attend such functions, with the exception of CEO Terry Strachan, but in their own way continued to provide an essential service to the RAS. At the end of 2011 Dee Newton left the Society's employ and was replaced temporarily by Sanele Khwela. In 2012 Priscilla Pandither was appointed as full-time use of grounds co-ordinator.

That year Iona Stewart's term as the Society's first female president (2010–2012) came to an end and in 2015 the cattle or B arena was renamed the Iona Stewart Arena in recognition of her contribution to the Cattle Section.[80] She was succeeded by Mike Moncur with Kay Makan and Tim Nixon serving as his vice-presidents.

Kay Makan became a vice-president of the RAS in 2010 after joining the executive committee in 2003. He was the only member of the Society's executive not drawn from the white ethnic group other than those who from time to time represented the City Council. After matriculating at Sastri College in Durban he had attained his GCE O and A levels in London where in 1967 he also acquired a diploma in electrical engineering at the London Polytechnic.

On his return home Makan pursued a business career and in 1991 established Kay Makan Electronics, followed by other stores in KwaZulu-Natal. He served terms as president of the local Chamber of Commerce, the Community Chest and the Gujarati Vedic Society among numerous other activities, as well as receiving several awards for business leadership and support of philanthropic causes. He initially joined the Society as an exhibitor at the Royal Show before Ron McDonald encouraged him to join the RAS executive. While subsequently undertaking to serve as vice-president he had no interest in eventually becoming the Society's president having already assumed that responsibility in other bodies.[81]

In 2012, after a prolonged adjudication process, the RAS was recognised as a level-two BBEE contributor. This was re-affirmed for the next five years and was significant in developing business opportunities. As anticipated, amended BBEE codes subsequently led to a downgrade from level two to five which was still above the national average.

It was hoped by way of more staff training to improve this to level four. In 2016 the Society was again deemed to be level-2 compliant and resolved to spend R40 000 on further skills development in anticipation of further regulatory changes. By the end of 2017 the RAS was tying with City Lodge

Hotels in 35th place among the JSE's 63 most empowered companies.

In 2012, wage negotiations with NEHAWU were successfully concluded with a 9% increase that was 3% above the current CPIX. In 2014, a CCMA intervention was necessary when the union, representing the Society's hourly paid staff, declared a dispute in connection with minimum wages thereby extending the annual negotiations. Cordial relations were soon re-established and maintained with the union. Subsequent wage negotiations, such as a 6.5% adjustment in 2018, were conducted satisfactorily.

In 2011–2012 two members of the administrative staff had to be summarily dismissed for 'indiscretions and a fraud'. In one case R32 000 in misappropriated funds were recovered from a personal provident fund.[82] Two longstanding and loyal grounds staff employees, Ncamsile Ngidi and Elphas Ngidi, had to be boarded due to ill health. In 2014 the most senior grounds employee Dumisani Ngcobo died, exhibitions co-ordinator Carmen Nänni left after nearly a decade of service to be replaced by Lara Bezuidenhout, and Eunice Banda retired after 25 years of overseeing cleaning and the provision of refreshments.

In 2014 the implementation of 'staff key performance indicators' was initiated under the mentorship of Neville Thomas. This required staff members to codify their tasks and intentions so that their performance could be measured for subsequent increments and bonuses. It was soon extended and well established, involving all members of the salaried staff.[83] That year Rochelle Naicker joined the front office staff but resigned in 2015. Margaret Mitchell was appointed temporarily before Janice Will assumed the role of secretary/personal assistant to the CEO.

Her predecessor, the highly experienced livestock co-ordinator and secretary/personal assistant to the CEO Jenny Fraser (formerly Perry, formerly Van Niekerk), emigrated to Britain. She had been an invaluable asset to the Society's administration, not least the Horse and Cattle sections, since joining its staff in 1983. Mike Moncur, among others, acknowledged a huge debt of gratitude to her for passing on her extensive knowledge of the livestock sections. Indeed, before her departure he persuaded her to compile a large dossier of notes to which he subsequently referred for guidance as 'Jenny's Bible'.

She was grateful for the honorary life membership of the Society accorded to her in 2002 and believed that her life had been 'enriched immeasurably through association with the Royal Agricultural Society and all its personalities, from volunteers to paid stewards to committee members as well as permanent and contracted staff'. She recalled 'so many happy and rewarding memories of my time at Natal's Royal' and that 'the weeks of long hours in preparation were well worth the result of being part of a wonderful, prestigious organisation'.

In her place voluntary stalwarts Malcolm Stewart-Burger and his wife were appointed jointly as Royal Show cattle secretary/co-ordinator. The former had been in the industry for many years but following their withdrawal they were replaced by Sheep Section chairman Jonathan Tyler, assisted by Andrew Adams and Petra Theron. When Tyler, in turn, withdrew Jenny Fraser returned temporarily as cattle secretary, being on site from 1 March to 30 June. She was assisted by Emmie Barnard and Andrew Adams, when available.[84]

In 2016 Dix Reddy, deputy grounds superintendent, opted to resign after 39 years of service. The following year grounds staff member T. Cele was given a cash award for 35 years in harness, Wendy Burnard was appointed on contract to replace the indisposed Barbara Shaw, and Julia Hackland became livestock secretary with Jenny Fraser undertaking to mentor her at long distance. The Society's constitution was updated by excising various provisions which had fallen into disuse and the office of RAS trustee was dispensed with.

Paula Greyling joined the staff as bookkeeper in 2017 when Barbara Shaw subsequently retired after fourteen years of service but, like Margaret Mitchell, the latter continued to assist with specific projects such as overseeing livestock prize monies, the asset register and BBEE regulations. That year the Society committed R50 000 to formal training programmes for deserving staff members.[85]

In 2018 the obligatory ten-yearly registered name Royal Show was officially confirmed. That year it was resolved to reduce the front office staff from three to two due to poor out-of-show trading conditions. In 2018–2019 the RAS was awarded a level-4 BBEE rating in terms of the new codes putting it in the top 35% of the country's businesses.

Following the appointment of three successive bookkeepers, Elizabeth Gaybba was appointed to that office. She joined an administrative staff which by then, in addition to Janice Will (personal assistant to the CEO), Lara Bezuidenhout (exhibitions manager) and Priscilla Pandither (venue hire manager), included Irene Peters (private functions co-ordinator), N. Mapstone (assistant venue hire co-ordinator) and L. Storey (membership administrator). Roy Motilall remained in office as grounds superintendent, assisted by K. Naidoo as artisan.

In 2019 Lara Bezuidenhout left to take up a senior marketing post elsewhere and was replaced by Lauren Fisher. The decline being experienced in the conference and events industry unfortunately necessitated Priscilla Pandither being offered a half-day position which she declined in favour of a full retrenchment package.

While the future legal persona of the RAS was still unresolved, it was nevertheless decided to improve the Society's management and executive structures along the lines proposed by Oliver James of TMJ Attorneys who had been commissioned to present possible alternatives. This was in response to frequent failures to raise a quorum and in the interests of avoiding unnecessary repetition.

The RAS had streamlined its administration once before, in October 1995, largely under the leadership of Derek Spencer. It was now proposed that in place of the existing meetings which were 'cumbersome, repetitive' and 'poorly attended', the executive committee should operate like the board of directors in a company. The CEO, who would continue to be responsible for overseeing operational matters, would be answerable to the board which, in turn, would be answerable to the Society's members.

It was envisaged that the board chairman/president and/or CEO would report to them at annual or special general meetings. The board would have fewer members (ten at most) and meetings, which could be held electronically, but would have the authority to create sub-committees and consider their recommendations. The 'plethora' of honorary office bearers would be 'pruned' and board members regularly rolled over without the need for annual re-election.

The executive committee unanimously approved these proposals which were also approved at the following AGM in November 2018. They had the desired effect with an executive no larger than eight by 2019, but without a female presence.[86]

Stalwart supporters and volunteers, like the RAS staff, also continued to be vital to the ongoing success of the Royal Show and the Society's other activities. Former president Derek Spencer (1991–1995) echoed the views of many others in gratefully remembering 'the many volunteers who give so generously of their time and the few paid but dedicated members of the permanent staff'. In some sections, like Crafts and Home Industries, the average age of volunteer workers was noticeably increasing.[87]

In October 2011, in recognition of their longstanding contributions, a lunch was held in honour of two popular Showgrounds personalities. Former vice-president and current honorary life vice-president George Poole and his close friend of many years, electrician Ken Easthorpe, were both celebrating their 80th birthdays. Poole died two months later.

Ken Easthorpe's long association with the Showgrounds began in the mid-1950s when he travelled from Durban to play badminton in the Olympia Hall. He subsequently secured temporary contract work for the RAS which, after the 1957 Show, became permanent. He was involved in the modernisation of the Showgrounds, including the installation of sealed time switches that for 44 years effected a significant saving in electrical costs. He was briefly banned from the grounds for mischievously soaking the grounds manager in front of invited guests by switching on the power while he was adjusting the rose of a newly installed fountain!

Easthorpe developed an interest in designing exhibition stands while working for the Sugar Association and for 20 years constructed its exhibits, several of them prizewinners, at the Show and in other centres. He was still responsible for the reticulation of the Showgrounds but his commitments further afield obliged him to decline the vacant post of grounds manager. He continued his association with the RAS after the closure of the Sugar Pavilion in the mid-1990s.

In 1975 Ken and Dawn Easthorpe held their wedding reception in the grill room and as both worked there the Showgrounds became their children's playground and other home. Mark Easthorpe eventually joined his father as a contract electrician while his brother Chris went into the poultry industry following a poultry show there. By 2019, after an association lasting '62 years and counting', Ken Easthorpe had survived many crises, electrical and otherwise, as well as injuries and had worked with thirteen RAS presidents, three general managers and one CEO. He died after a fall in 2020.[88]

In 2011 the RAS also mourned the loss of honorary life member Audrey Shepherd who had contributed so much to the Cattle and Equestrian sections and Ian Dixon, formerly of FNB and a longstanding supporter of the Society. Honorary life membership was awarded to Marion Harper, John Oliver and Jo Olley.

At the end of her term as president Iona Stewart was elected an honorary life president and was applauded for her sustained contribution to the Cattle Section over many years as well as her service to the executive since 2001. Also, for being what Ron McDonald called the Society's 'first lady' in her capacity as the first female president in its 160-year history.[89]

In 2013 the deaths of former president David Wing (1998–2003) and Dawn Meyer, convenor of the Ilovo floral icing competition, were recorded. Future president Mike Moncur was made an honorary life vice-president after serving on the executive for ten years and vacating the position of vice-president. Honorary life membership was awarded to longstanding stalwarts Bert and Barbara Cornell. Bert had been responsible for the public address system at the Showgrounds for 39 years although by then he was focusing more specifically upon the Royal and Garden shows.[90]

That year Kay Makan was another who joined the ranks of honorary life members after being on the executive for ten years and a vice-president since 2010. In 2013 honorary life membership was also accorded to Bob Mitchell, a steward and judge in the Cattle and Sheep sections for more than 35 years, Brian van der Bank, after more than 30 years' involvement in the livestock sections, Jacque Webb, who had been active in the Cattle Section in excess of 30 years and Keith Jones, a highly respected livestock judge and steward of longstanding, who died soon afterwards.

The Society also lost Jamie Main-Baillie who had overseen much of the Main Arena entertainment and Bertram Mapstone who for more than three decades had exhibited his Sussex cattle at the Show. In 2014 Alan Shaw and C.A. Froneman, who had both served on the executive for ten years, were accorded honorary life membership. So too was Garth Carpenter in 2015 after being associated with the RAS for 44 years presenting his educational reptile exhibit.

Lieutenant Jack Haskins was given that recognition a year later. He was shortly to retire from the SAPS after a career-long contribution to the community and prominent participation in the Hall of Safety and Security at the Show. That year honorary life vice-president Bruce Lobban died as did Dallas Kemp, who was remembered as 'a well-liked and knowledgeable contributor to the bovine section' and Rob Hewitt, who for a decade had overseen the Royal Shows' dog performances. Moira Crookes, former cattle committee member, died in 2016.[91]

The following year the deaths of several more of the Society's stalwarts were recorded. They included Ann Duckworth, secretary of the Crafts and Home Industries Section, Trevor Marwick, husband of Lynn Marwick an honorary life member and pillar of the Cattle Section, Arlene McDonald, wife of former president Ron McDonald, Harry Tully, an honorary life member and pig and sheep committee member for more than 50 years, Rowly Waller, former general manager (1994–1996) and Una Watson, member of the floriate and Garden Show committee.

Stalwart supporters whose deaths were recorded in 2018 included David

Glaister, longstanding supplier to the RAS, Dr Nancy Rayner, an active contributor to the Crafts and Home Industries Section for 68 years, Joe Toth, chief steward of the Rabbit Section for more than 30 years, and Eddie Peters, husband of Irene Peters who worked in the administration office.

Dr Raymond Nixon, veterinarian and longstanding member, also died as did David Davidson, the designer for many years of South Africa's annual Chelsea exhibit and latterly the Garden Show's senior judge. Bobby Hex, valued member of both the executive and management committees, emigrated to Australia. In 2018–2019 two former vice-presidents, John Plummer and Pieter Breytenbach, passed away as did Ian Cameron who for 25 years had overseen all Showgrounds stage construction.[92]

Security at the Showgrounds was always a concern in the interests of staff, volunteer assistants and visitors as well as the Society's own property. Unfortunately, there were ongoing periodic thefts of cellphones, laptops and purses, more particularly during Show times.

In 2012, to improve security and streamline the issue of access permits, the self-seal ID pouch was re-introduced. There were nocturnal pranks within the Showgrounds when student assistants were allowed to sleep overnight in cattle clubhouses and supplied over-generously with alcohol. An implement exhibitor complained that a tractor had been removed from his stand overnight to the cattle area![93]

Two of the Society's valuable silver floating trophies disappeared in 2013–2014 but mysteriously re-appeared during the 2014 Show, seemingly floating in more senses than one. This incident prompted the RAS to implement tighter security measures and to establish a photographic register of its trophies and record them digitally on its computer network. On another occasion the Royal Industrial Floating Trophy was stolen from an exhibitor's stand, but fortunately the thief was apprehended at the funfair with the large bowl-shaped prize squashed inside his jacket.[94]

In 2014 Superior Security became the new security provider at the Showgrounds after Imvula had performed the service for 20 years. Following several break-ins at various venues during 2015 it became necessary to begin installing an armed response system in some buildings. Most breed clubs opted to remove their valuable items outside of Show periods.[95] Petty thefts and occasional vehicle thefts were still ongoing. More seriously, all copper and metal conduits eventually had to be replaced with polycop.

In 2017 there were several more burglaries, particularly in Block B adjacent to the Victoria Siding, necessitating improved lighting and security. Occusec now assumed responsibility for security and, to improve the accuracy of the foot count and eliminate scams, the decision was taken to replace the identity cards issued to exhibitors and members with biometric (fingerprint recognition) access. The 5% of older members with feint fingerprints were issued with magnetic cards. Hopefully, they all resisted the temptation following this discovery to embark upon a new career in burglary!

Unfortunately, slow registrations caused difficulties immediately prior to and during the first two days of the Show. The system was used with some success at the 2017 Garden Show but there were subsequent difficulties with fingerprint readers at the 2018 Royal Show. It was decided that in future fingerprints would no longer be required, exhibitors would gain access to the Showgrounds via a ticket allocation while permanent operators on site and members would do so via a magnetic swipe card system that would also serve as staff and membership identification. No further difficulties were encountered.

In 2018 CCTV was installed along Sharks Alley and at gates 3A and 7. The east bank of the Chatterton Road boundary had to be secured to prevent the flood control canal being used for washing taxis, with an attendant problem of litter. The installation of fencing and signage seemed to have the desired effect while the pathway to the other side of Chatterton Road through the Victoria Siding adjacent to Blackwoods Nursery was also securely fenced.[96]

Security upgrades were but one dimension of the physical improvements to the Showgrounds and facilities that were so essential to attracting exhibitors and visitors to the annual Royal and Garden shows as well as remaining competitive in successfully hosting a variety of other events.

Physical improvements in the years leading up to the 160th Royal Show in 2019 were to some extent limited by the loss of the Society's tax-exempt status which reduced the cash resources available for such purposes. Even so, in 2012, following extended negotiations with the municipality, the RAS agreed to bear the estimated R350 000 cost of installing the traffic light outside gate 1 provided it came under the Society's ownership. It was in place for the 2012 Show and reduced the need for a dozen or more traffic officers to one or two.

That year, following earlier concerns, Terry Strachan inspected all the Showgrounds facilities with expert Michael O'Flattery before they were 'broadly found to be disabled friendly'. In January 2013 more than R700 000 worth of damage was incurred when a wind storm destroyed the members' grandstand roof and adjacent office. Fortunately, this expense was covered by insurance but the members' facilities also required unrelated refurbishments.[97]

Main Arena tower lighting support systems and the Umgeni Walkway were refurbished and expensive unbudgeted leaks in the Honey Hall were repaired. A new concrete cattle bridge was constructed in place of the drift behind the EO Jones Pavilion with a plaque acknowledging the substantial input of deputy grounds superintendent Dix Reddy. By way of appropriate decoration, seven sheep figurines were added to the illuminated cattle images that had already been installed adjacent to Gate 1.

In 2013 the equestrian clubhouse was refurbished as were the administration offices with new downlighters and an air conditioning system. Halls 8 and 9 were also air conditioned but unbudgeted expenditure was necessary when the flood control canal burst its banks and inundated the cattle arena area to a depth of 30 cm.

After much deliberation it was decided not to proceed with the construction of a large R37 million multi-purpose hall as this was beyond the Society's means. However, the first phase of a new 700m^2 venue, Hall 10, was initiated in 2014 between the Bonsmara clubhouse and Hall 9 following the demolition of the old cattle stall blocks on that site. It was the Society's 'first truly multi-purpose venue catering for both livestock and general functions' and was 'in line with Royal Shows elsewhere in the world'. It was used for the first time to house cattle but while its value was fully appreciated its completion had to be delayed due to the expense involved in the current economic climate.

Nevertheless, serious consideration was given to installing a 'photovoltaic energy system' by installing solar panels at the Showgrounds for about R2 million. It was anticipated that, outside Show periods, it would provide 100% of the Society's electrical requirements, would cover roughly 20% of its annual electricity bill and pay for itself in thirteen years. The expenditure was not approved, but in April 2015 it was suggested that the RAS might reconsider it if Pietermaritzburg followed the eThekwini municipality's example by offering to remunerate private photovoltaic energy providers for electricity fed into the national grid.[98]

That year the former Natal Parks Board stand was divided into two and the rare cycads there were re-planted in front of the Olympia Hall. R130 000 was spent on refurbishing the reception area of the administrative office for the first time in 60 years while the computer server and photocopier were replaced for R160 000. Gabion baskets were used to form a quasi-dam in the Dorpspruit adjacent to University House to filter the river and encourage bird and other wildlife. A hailstorm in February inflicted R1 700 00 in damage but insurance paid for roofing and grandstand seating, some of which needed refurbishment anyway.

In 2016–2017 Santam contributed a generous R720 000 towards further refurbishment of the Showgrounds roofing and there were improvements to other facilities including the Royal Food Court (previously Scott's Catering), with Gary Camps as caterer. Shortly afterwards Terry Strachan had an unusual experience when ordering a bottle of wine for his colleagues. On arrival at his showtime table, strategically placed on the Main Arena perimeter, the waiter opened and sampled the contents himself before offering his glass to Strachan for confirmation that it was drinkable![99]

There were also numerous maintenance costs including a revamp of Total Tower in the Cattle Section and of the cattle-handling facilities, the reinforcement of the north bank of the Dorpspruit with gabions and the removal of the Umgeni Walkway east of Dhodas Bridge due to its irreparable degradation. In 2016 the Olympia Hall was revamped to host the OMI anniversary.

Halls 7 and 8 were subsequently joined to create a single, more versatile 1 500m^2 venue. By mid-2016 the Society's WiFi system extended over all of these facilities, including the administrative offices, with further reach planned.[100] In February 2018, the server for the whole computer base crashed, necessitating its expensive replacement.

Later that year, as a grim reminder of earlier crises, heavy rains breached the flood control canal depositing mud in the vicinity of Gate 5. Another four-yearly excavation of the canal proved necessary. In addition, a safer and more user-friendly Olympia Hall entrance was completed at Dhodas Bridge.[101]

Despite the prevailing financial constraints many improvements had been made to the Showgrounds and much water had flowed down the Dorpspruit, mostly effectively controlled, by the time of the Society's 160th Royal Show in 2019.

The **160th Royal Show** like the 125th with which the modern era of the RAS had begun in 1984, was similarly a great success favoured with good weather despite the disappointing decline in attendance. As always, the intention was, as Terry Strachan put it, to introduce 'town to country' and provide the public with 'an educational and fun-filled experience' at what was still 'South Africa's largest mixed agricultural show in terms of livestock and equipment'. Some 350 visitors from the greater Durban area took the option of making a return train journey from Kloof station to the Showgrounds.

Attendance was also encouraged by implementing the previously successful Super Friday when caterers, exhibitors and the funfair offered special discounts.

The live broadcast by RSG of its 'Brekfis met Derrich' programme was attended by 130 guests and reached more than a million listeners while 200 invitees enjoyed the farmers' braai in the cattle arena. Fiso Hadebe, winner of the Young Auctioneer senior competition that year, oversaw a novelty auction at the Gold Cup dinner which raised R11 000 for the SPCA.[102]

Main Arena displays featured a return of the Puma Flying Lions with their Harvard aerobatic performance as well as sky diving and yet another Royal Demolition Derby involving competing professional stock car racers. There was also the regional leg of the Powasol Enduro-X champion series and a spectacular Monster Night Fight Show performed by freestyle motocross stunt riders on a giant rig combined with pyro flames, FMX trains and soundtrack.

In addition to a strongman display Husqvarna's Bring out the Manne involved axe throwing, log sawing, sawdust clearing and a mini ride-on-race. Among more sedate Main Arena events were the Gold Cup Parade of top-quality cattle, sheep and goats, the popular KwaZulu-Natal dog agility trials and lunchtime Working on Fire demonstrations.

Additional attractions included pupils from local schools performing in a Rock 'n Royal Marimba Fest, the usual crowd-pulling Royal Symphony Concert featuring ever-popular waltzes and marches, the RSG Concert starring Francois Henning (aka Snotkop) and the legendary Sonja Heroldt, the Hindvani FM Extravaganza with performers from Aradhna's Dance Academy and the East Coast Radio Royal Rock Concert headed by the electro-swing dance group Good Luck.

The Show's musical attractions culminated in the Sisonke Festival on the final Sunday with a DJ competition and performances by 2018 South African Music Award-winner Sun-El Musician as well as local artists Simmy and rapper Nthuthoko Mkhize, aka Prodii-G, among others. Despite being well advertised in the local press attendance was disappointing.

The KFC River Stage provided family entertainment featuring the usual shark dissections and informative scientific experiments as well as clown, magic, mime and jazz performances, fitness demonstrations, a talk on wild animals, and a presentation of creepy crawlies and snakes. For the more physically active there was the traditional funfair, camel rides, a new artificial fibre skating rink capable of accommodating 60 skaters at a time, the Karkloof Canopy Zipline and the 18-metre freefall Bagjump into a giant inflatable mattress.[103]

The **Horse Section** retained its appeal with show jumping, dressage, tent pegging, children's and national showing classes and the Natal Boerperd championships. Enos Mafokate, the nation's first black show jumping champion, was in attendance.[104]

The **Cattle Section**, in former president Ron McDonald's estimation, was still 'the beating heart of the Show' with a pleasing 705 cattle entries at the 2019 Royal Show compared with 662 in 2018 despite the prevailing tendency towards on-farm assessment and the absence of dairy cattle.

Among other highlights the Show included the Dexter national championships. Frank Hinze, owner of MiniMoo Dexters in Howick and head of breed improvement for Dexter SA, affirmed that with its tradition and culture as the oldest agricultural show in the country, it 'presents a classy platform with great media coverage for our society to promote our breed to the public'.

The first weekend featured young up-and-coming farmers displaying their knowledge and handling ability of cattle and sheep in the 38th KZN Youth Show and the Future Farmers competition. The Living Land Workshop first introduced in 2017 involved specialists presenting thirteen modules to 200 emerging farmers. Youth also featured in the KwaZulu-Natal leg of the Toyota SA National Young Auctioneer competition which comprised two categories: juniors who were 19 years old and younger, which Fernando Craighead won; and those between 20 and 35 years of age, won by Fiso Hadebe.

For the first time in seventeen years Weston Agricultural College won the J.R. Brewitt Floating Trophy for the best string of four halter-trained animals. Weston pupil Sven Lindeque won first prize in the 450 kg and over class in the steers competition and Ayron Chatten won the 400 kg and under category. Manqoba Mafuleka of the Mtubatuba Mini Farm was successful in the 401 to 450 kg group.

Andrew Masterson's Black Angus Milagro Quebec 1704 was the supreme champion beef bull while Colette and Amy Masterson's Simmentaler Milagro Amy 1471 was the supreme champion beef cow. A regular participant based in Humansdorp, it was Masterson's fourth Royal Show Gold Cup and in his estimation was 'the one we want to win because of all its incredible tradition, the fanfare and everything that goes with it'.

John and Tracey Devonport's Limousin Devlan Hearsay 1266 was the reserve supreme champion bull and Gawie Naude's Braunvieh from the Eduan Boerdery in Barkly East was the reserve supreme champion beef cow. Dr Robert Kleinloog's Netherwood Big Mama from Nottingham Road's Netherwood Stud was named the senior and reserve grand champion cow in the Angus competition.[105]

The **Sheep Section** yet again hosted the Hampshire Down national championships. Among the exhibitors was Russell Shorten who had been competing since 1965. There were also regional championships for Dormer, Dorper, Ile de France and Suffolk sheep as well as Boergoats and a Veldgoat sale. A Royal Show record for small stock entries was established with the presence of 1 076 sheep, lambs and goats. Another attraction was the ever-popular Natural Wool and Fibre Expo, which included the Masibumbane sewing group from Mpophomeni who offered a variety of items for sale.

The supreme champion ewe and ram at the Show were both White Dorpers exhibited by Dr Corrie Avenant from the Aka Stud in Williston. The reserve supreme champion ewe was Regina Harmse's Ile de France from Ermelo and the reserve supreme champion ram a Suffolk from the Debak Suffolk Stud owned by Hennie Geldenhuys from Memel. Dr Avenant's Boergoat rams and ewes, at the Royal Show for the first time, were also the supreme and reserve supreme champions in their categories. It was the first time that one exhibitor had swept the boards in both the sheep and goat categories.[106]

The **Carcass competitions**, now in their 28th year, were livestreamed on Facebook to 35 000 people nationwide and attended by more than 200 people, with 98 cattle, 88 lambs and 169 pigs entered. They included the sale of a top-grade Wagyu carcass, a first in RAS history but not included in the SAMIC competition. Johan du Plessis produced the 420 kg animal that was sold for R160 000 to Zingela Meats which achieved another first by bidding on the phone, ahead of others from all over the country.

One of four Japanese beef breeds and rated in some quarters as the best quality meat in the world for its marbling, tenderness and taste, the Wagyu Society of South Africa sought to promote it with its stand at the Cattle Expo and stock of 1 600 burgers which were sold out within three days.

The Beefmaster, which Clark Rattray of Pleasant View Beefmasters in the Swartberg bred, set a new national record when it was sold to regular buyer the Oyster Box Hotel for R200 per kg compared with the ruling price for beef of R45. Tahir Kharwa of Weston Agricultural College was responsible for preparing the animal which weighed 271.1 kgs and achieved a SAMIC score of 94,54%. The first reserve champion beef carcass was a Bonsmara from Mutubatuba Mini Farm that the Oyster Box bought for R100 per kg.

The champion group of beef carcasses were also Bonsmaras, which the Vryheid Landbou High School had bred and prepared while Weston prepared the first reserve champion group of Beefmasters. Richard Tedder produced the champion super ox carcass, a Brangus which sold for R45 per kg.

Regular exhibitor Gert Lotter from Hofmeyr maintained the Show's tradition of attracting good lamb prices when his Ile de France/White Dorper cross scored 97.5% and sold for R1 500 per kg to the Oyster Box compared to the ruling price of R665. Lotter also produced the 27.5 kg champion super lamb carcass while E. le Roux bred the champion group of White Dorpers and Regina Harmse produced the champion and reserve champion lamb carcass with European genetics.

Glenelly owned the champion carcass, a PIC, in the pork section and sold it for R90 per kg to the Oyster Box compared to the current ruling price of R21. Farmhouse Butchery bought B. Gumede's PIC emerging farmer carcasses at the latter price and Taylor's Meats in Ballito also acquired the New Hanover Evangelical Lutheran Church's champion group for R21 per kg. Koos van der Ryst of the RPO congratulated all concerned on the high standard of the carcass competitions and the quality of the animals on display, with a new South African record of R200 per kg being set for the sale of a beef carcass.[107]

Other agricultural sections, as before, provided the province's largest combined display of feather and fur in the form of birds, rabbits and apiarian products as well as the usual variety of agricultural implements. The Pietermaritzburg Canary and Cage Bird Club celebrated its 107th birthday with its annual open bird show attracting exhibitors from all over the country. The local Budgerigar Club hosted the KwaZulu-Natal provincial show on behalf of the Association of Wild-Type and Exhibition Budgies of South Africa which also attracted breeders from far afield.

So, too, did the competitions held by the Natal and Coast Poultry Club and the Natal Rabbit Club with the latter hosting British judge Peter Faint and attracting 343 entries from 29 exhibitors. There was a decline in entries in the Apiarian (Honey) Section but previous high standards were maintained. Reg and Kim McCall and the Solomon family won the trophies and there was a noticeable revival of interest in the Mead classes and novelty wax categories.[108]

The **Crafts and Home Industries Section**, as in previous years, showcased a wide variety of the hobbies pursued in the region. In particular, it promoted the work of the Midlands Wood Workers Guild whose members displayed their skills with a series of demonstrations that attracted public interest. They won the Special Endeavours Award while their chairperson Aubrey du Plessis secured the Austin Smith Memorial Floating Trophy.

The hall as a whole was awarded a gold medallion and the President's Cup for Special Endeavour while June Henstock was the overall champion with Janet McCoy the runner-up. Nicola-Paige Hanegraaf was the junior champion for the third consecutive year and Swartkop Valley the Women's

Institute winners. Also present were Tracy Cull, Fleur and Michelle Robinson, founders of the online shopping centre, which offered numerous homemade goods ranging from leather work, jewellery and mosaics to beaded bowls and stained glass.[109]

The **Commerce and Industries Section** boasted 478 exhibits, including service sector displays which featured those of the SANDF and SAPS among others. The Mercury Food and Festival of Fine Living was as popular as ever while the Daily News Hall focused on fashion ware. Umgeni Water provided free classes for schoolchildren with the emphasis on water purification processes and the importance of saving water.

Among the numerous prizewinners were Midlands Mascor, which secured the Hulamin Floating Trophy for best display on an open site and the Pietermaritzburg Chamber of Commerce Agriquip Floating Trophy for the best display of heavy-duty agricultural equipment. Cherylin's Creations was also a double winner being awarded the Royal Commercial Floating Trophy for the most outstanding display in an exhibition hall and the RAS Trophy for special endeavour.[110]

The overall success of the 2019 Royal Show and favourable publicity that it received confirmed that it still had relevance to exhibitors and to the general viewing public. However, at least one visitor expressed regret at the relative absence of indigenous crafts, along with that of nGuni cattle. He also pointed out that there were no exhibits from 41 of the 43 local municipalities in the province. It therefore remained, in his view, 'a very Western style show'. Unfortunately, although he had avowedly spent sixteen years 'working with rural people in KZN from Pongola to Umzimkulu', he offered no insights as to how these defects might be overcome.[111]

The existence of such inadequacies was certainly not for the want of seeking solutions over many years, more so by some sections of the Show than others. According to RAS president Mike Moncur, nGuni cattle owners seldom, if ever, showed their livestock. Moreover, repeated efforts to include indigenous sheep by offering various incentives to breeders via the provincial Department of Agriculture and extension officers had failed. Attempts to attract local craftsmen had similarly been unsuccessful and raised issues with regard to providing transport as well as accommodation for the duration of the Show.[112]

These shortcomings pointed to but one of many challenges still facing the Society. In the light of these and gathering uncertainties it remained to be seen whether and in what form the RAS and its various activities would survive in the future.

ENDNOTES

1 This chapter draws on *AR*s 2012–2019.

2 *AR* 2012: 22.

3 *AR* 2017: 24.

4 Mike Moncur, interview, 26 June 2019.

5 RASM, Executive, 2 November 2011: 3.

6 RASM, Management, 11 September 2013: 3; Executive, 17 September 2013: 4.

7 RASM, Executive, 22 April 2014: 4.

8 RASM, CEO's Report, 11 August 2015: 2 and 20 July 2016: 3.

9 *AR* 2017: 15. RASM, CEO's Report, 8 November 2017: 2.

10 RASM, CEO's Report, 21 February 2018: 3. *AR* 2019: 10, 16.

11 RAS, *Prospectus 2012–2015*: 7. RASM, Agricultural Services, 12 July 2013: 1; Management, 25 April 2017: 2 and 27 June 2017: 2. *AR* 2019: 4, 6, 7, 10, 31.

12 RASM, AGM, 4 November 2015: 4.

13 *AR* 2019: 5, 22, 31.

14 John Fowler, interview, 10 June 2019. RASM, AGM, 2 November 2011: 4; Agricultural Services, 20 July 2012: 2.

15 Mike Moncur, interview, 26 June 2019. RASM, CEO's Report, 17 July 2013: 3.

16 Iona Stewart Personal Papers, 'Beginning of it all'.

17 RASM, Executive, 24 June 2014: 3, 4; CEO's Report, 29 June 2015: 2.

18 *AR* 2017: 9, 2018: 10.

19 RASM, Management, 18 January 2012: 1; CEO's Report, 18 January 2012: 2.

20 *AR* 2012: 13.

21 RASM, Cattle, 7 September 2013: 1.

22 RASM, Executive, 21 January 2014: 2; Cattle, 29 January 2014: 3 and 23 July 2014: 2.

23 RASM, Cattle, 4 November 2015: 2; CEO's Report, 16 March 2016: 3.

24 RASM, Cattle, 22 June 2016: 7 and 16 November 2016: 6; Executive, 17 January 2017: 2; Management, 15 February 2017: 3, 27 June 2017: 1, 16 January 2018: 2–3 and 13 March 2018: 2.

25 RASM, Executive, 27 June 2017: 8; Sheep and Goats, 28 June 2017: 1; Management, 16 January 2016: 1–2.

26 RASM, Cattle, 14 March 2018: 2; Executive, 26 June 2018: 3.

27 *Witness*, 5 January 2018.

28 RASM, Management, 24 June 2014: 3.

29 RASM, Executive, 22 January 2013: 4. *AR* 2016: 9.

30 *AR* 2018: 8.

31 RASM, CEO's Report on 2018 Royal Show, 25 June 2018: 1–3.

32 RASM, Cattle, 20 February 2013: 3. *AR* 2013: 13.

33 Mike Moncur, interview, 26 June 2019.

34 *AR* 2018: 12–13.

35 RASM, CEO's Report, 15 August 2018: 3; Management, 15 August: 2.

36 Tim Nixon, interview, 1 August 2019.

37 *AR* 2017: 14.

38 RASM, Agricultural Services, 9 March 2012: 1; CEO's Report, 29 June 2015: 3 and 17 November 2015: 3; *Farmer's Weekly*, 12 July 2013.

39 Mike Moncur, interview, 26 June 2019. RASM, CEO's Report, 18 January 2012: 2. *AR* 2013: 16.

40 RASM, Executive, 17 September 2013: 5 and 19 January 2016: 4.

41 RASM, Executive, 17 January 2017: 5.

42 RASM, Executive, 26 June 2012: 4.

43 *AR* 2016: 17.

44 Di Fitzsimons, interview, 4 June 2019.

45 RASM, CEO's Report on 2012 Royal Show, 25 June 2012: 2.

46 RASM, Commerce and Industries, 13 March 2012: 1.

47 *AR* 2014: 9.

48 RASM, CEO's Reports on Garden and Leisure Shows 2 October 2012: 1–3, 23 August 2013: 1 and 13 November 2013: 3; Executive, 17 September 2013: 5; Management, 20 November 2013: 2; Sunday Tribune Garden Show, 25 September 2014: 1–3.

49 Terry Strachan, interview, 12 March 2019. RASM, Management, 12 November 2014: 2 and 18 March 2015: 2; Executive, 22 April 2015: 5.

50 https://www.tanya.visser.com.

51 RAS, *Prospectus 2012–2015*: 12–13; RASM, Management, 18 November 2015: 2.

52 *AR* 2015: 19.

53 Terry Strachan, interview, 12 March 2019. RASM, CEO's Report on Witness Garden Show, 12 November 2015: 1–5; Reports, 9 November 2016: 1, 24 October 2017: 1–4, 8 November 2017: 1 and 16 October 2018: 3; Management, 16 January 2018: 3. *AR* 2017: 21, 2019: 17–20. *Maritzburg Fever*, 12 September 2018. *Witness*, Garden Show Supplement, 4, 6, 7, 9 September 2019. *AR* 2020: 5, 10–15.

54 RASM, CEO's Report, 16 October 2012: 2; Executive, 23 April 2013: 4. *AR* 2020: 7–9.

55 *AR* 2013: 7, 2019: 16. Iona Stewart Personal Papers, 'Miscellaneous anecdotes', n.d. RASM, CEO's Report, 9 September 2014: 2.

56 RAS, Office of the CEO, 'The Royal Agricultural Society of Natal strategic overview' (unpublished typescript, January 2016): 6. RASM, CEO's Report, 11 August 2015: 2 and 16 March 2016: 2.

57 RAS, *Prospectus 2012–2015*: 8–10.

58 RASM, CEO's Report, 18 July 2018: 2 and 15 August 2018: 2. *AR* 2019: 15, 16, 21, 2020: 15.

59 RASM, Executive, 2 November 2011: 2; CEO's Report, 17 July 2013: 2, 13 August 2013: 2 and 18 March 2014: 3.

60 *AR* 2015: 24; RAS, Office of the CEO, 'Strategic overview': 9–26, RASM. Executive, 10 February 2016: 2–3 and 19 April 2016: 5.

61 RAS, Office of the CEO, 'The Royal Agricultural Society of Natal strategic overview': 8.

62 RASM, Executive, 28 June 2016: 5.

63 RASM, Executive, 12 September 2017: 2, 8–9; Management, 18 October 2017: 3, 16 January 2018: 2, 4, 17 April 2018: 2 and 18 July 2018: 3. *AR* 2019: 24–25.

64 RASM, Management, 16 January 2018: 6.

65 RASM, CEO's Report, 13 March 2018: 2, 3 and 16 October 2018: 2. *AR* 2019: 5.

66 RASM, Management, 15 August 2018: 3 and 25 September 2018: 2; CEO's Report, 15

August 2018: 2, 3 and 25 September 2018: 3. *AR* 2019: 6, 21.

67 *AR* 2019: 4–5, 24–25.

68 *AR* 2014: 16.

69 RASM, CEO's Report, 18 July 2018: 2. *AR* 2019: 17, 26.

70 RASM, Management, 14 March 2012: 2 and 18 April 2012: 2.

71 RASM, Management, 18 January 2012: 2–3, 15 February 2012: 2, 26 June 2012: 1–2, 16 January 2013: 2, 25 June 2013: 2 and 11 September 2013: 1–2; CEO's Report, 12 March 2013: 4; Executive, 17 September 2013: 2.

72 *AR* 2012: 22. RAS, Office of the CEO, 'The Royal Agricultural Society of Natal strategic overview': 7.

73 RASM, CEO's Report, 18 April 2012: 2–3 and 22 April 2014: 3; Management, 25 June 2013: 1 and 19 March 2014: 1.

74 RASM, CEO's Report, 12 September 2017: 2, 4; Management, 15 August 2018: 1. *AR* 2019: 22.

75 RASM, Management, 24 June 2014: 1, 13 September 2016: 2 and 12 September 2017: 2. *AR* 2019: 22.

76 RASM, Management, 18 January 2012: 3, 15 February 2012: 3, 19 March 2014: 3, 16 March 2016: 3 and 14 March 2017: 3; CEO's Report, 10 September 2012: 4 and 20 January 2014: 3–4.

77 RASM, CEO's Report, 10 September 2012: 3. *AR* 2013: 7.

78 RASM, Management, 9 September 2014: 2; CEO's Report, 11 September 2015: 2 and 13 September 2016: 4.

79 RASM, CEO's Report, 18 July 2018: 3 and 15 August 2018: 3. *AR* 2019: 12–13, 2020: 4, 6, 7.

80 RASM, Executive, 7 November 2012: 2, 5 November 2014: 2; Management, 18 March 2015: 3.

81 Kay Makan, interview, 2 July 2019. *Public Eye*, 29 March 2013.

82 RASM, Management, 18 January 2012: 2, 15 February 2012: 2, 14 March 2012: 2 and 19 April 2016: 3; CEO's Report, 18 January 2012: 2, 15 February 2012: 2 and 8 November 2017: 1. *AR* 2012: 10.

83 RASM, CEO's Report, 15 July 2014: 2 and 11 August 2015: 2; Executive, 9 September 2014: 3.

84 Iona Stewart Personal Papers, 'Note on Jenny Fraser', n.d. Mike Moncur, interview, 26 June 2019. Jenny Fraser, e-mail, 1 July 2019. RASM, CEO's Report, 16 October 2013: 2, 31 May 2014: 5 and 23 August 2014: 2; Management, 15 October 2014: 3; AGM, 5 November 2014: 4; Executive, 20 January 2015: 5; Cattle, 24 June 2015: 4.

85 RASM, Management, 19 July 2017: 3; CEO's Report, 19 July 2017: 1 and 16 August 2017: 2.

86 RASM, Management, 15 August 2018: 3; Executive, 25 September 2018: 2–3. *AR* 2018: 3, 2019: 2, 4, 5, 23, 2020: 18.

87 Iona Stewart Personal Papers, e-mail from Derek Spencer to Iona Stewart, 24 November 2015. Di Fitzsimons, interview, 4 June 2019.

88 Ken and Mark Easthorpe, interview, 27 August 2019. Ken Easthorpe, '62 years': 1–10.

89 RASM, AGM, 7 November 2012: 2.

90 Bert Cornell, interview, 18 June 2019.

91   RASM, CEO's Report, 16 November 2011: 3, 19 January 2016: 3 and 17 August 2016: 2; Executive, 23 April 2013: 5–6; Management, 14 August 2013: 3, 22 August 2014: 3 and 10 February 2016: 4. *AR* 2016: 23.

92   *AR* 2019: 22.

93   RASM, Cattle, 29 August 2012: 4.

94   RASM, Executive, 22 April 2014: 2; Management, 24 June 2014: 1. Jenny Fraser, e-mail, 1 July 2019.

95   RASM, Executive, 17 September 2013: 6; CEO's Report, 17 November 2015: 2; Cattle, 24 February 2016: 3.

96   RASM, Executive, 25 April 2017: 3; Management, 18 October 2017: 1, 16 January 2018: 2, 26 June 2018: 1, 26 June 2018: 3, 18 July 2018: 1 and 25 September 2018: 4.

97   RASM, Management, 14 March 2012: 2, 18 April 2012: 1; Executive, 26 June 2012: 3; CEO's Report, 16 January 2013: 3.

98   *AR* 2014: 5, 21, 2015: 7. RASM, Management, 18 April 2012: 2, 25 June 2013: 1, 21 January 2014: 1 and 22 April 2015: 1; Executive, 21 January 2014: 5.

99   Terry Strachan, interview, 12 March 2019.

100 RASM, CEO's Report, 28 June 2016: 2.

101 RASM, CEO's Report, 13 March 2018: 2 and 17 April 2018: 2. *AR* 2020: 5.

102 *Witness*, Supplement 'The Royal Show 24 May–2 June 2019': 2–3. *AR* 2019: 8–9, 14.

103 *Witness*, 23 May 2019: 2, 31 May 2019: 3. *AR* 2019: 9.

104 *AR* 2019: 15–16.

105 *Witness*, 24 May 2019: 8, 27 May 2019: 6, 30 May 2019: 8, 31 May 2019: 6. *AR* 2019: 11–13, 27.

106 *Witness*, 24 May 2019: 8, 27 May 2019: 6–7, 31 May 2019: 6. *AR* 2019: 6, 11, 27–28.

107 *Witness*, 29 May 2019: 6. Mike Moncur, interview, 26 June 2019. *AR* 2019: 12–13, 27–28.

108 *AR* 2019: 7, 15.

109 *Witness*, 27 May 2019: 7, 30 May: 8. *AR* 2019: 16.

110 *AR* 2019: 7–8, 28–30.

111 *Witness*, 7 June 2019: 9. *AR* 2020: 5.

112 Mike Moncur, interview, 26 June 2019.

5 RETROSPECT AND PROSPECT

BY THE END OF 2019, with 160 Royal Shows successfully completed, the RAS could look back with some satisfaction on its own career and forward with some confidence, if uncertainty, about its survival in the foreseeable future.

Retrospect

Beginning with its very first Show in 1851 the Society had overcome the vagaries of the weather, including violent storms, flood, hail, wind and drought despite so many of its activities necessarily taking place outdoors. It had come to terms not only with natural disaster but also with human tragedy. These included the 1984 Main Arena death of Mexican highwire artist Chuchin at the 125th Royal Show, the loss of a four-year-old child in a fire at the Chatterton Road stables in 2001 and the fatal accident of a six-year-old at the funfair in 2013, the first death of a visitor to the Showgrounds in 162 years.

The RAS had coped with financial crises and prolonged economic recessions, including those of the 1860s and early 1930s. Not the least of these had been experienced during the last decade, severely affecting the exhibition-based industry all over the world. Indeed, some similar organisations had been obliged to close, a notable example being the parent Royal Show of England.

Adverse economic conditions had invariably impacted unfavourably upon sponsorships. They also reduced the income derived from venue hiring when expenditure on non-essential events like conferences and exhibitions was trimmed from both governmental and private company budgets, including a substantial scaling down of the 2019 opening of the provincial legislature.[1]

The RAS had survived other financial setbacks, including the obligation to pay full municipal rates from 2003, with no softening rebate from City Hall. It was the only one among 53 similar organisations throughout the world that was required to do so. This seemed to reflect a growing indifference towards the Society on the part of local civic authorities.

There was also the change in SARS regulations which from 2012 obliged the RAS to pay income tax, even though it was a non-profit organisation with no individual benefiting from the surpluses it generated.[2] As a result there was less net annual income to spend on the maintenance, physical improvement and growing security needs of the Showground's aging facilities while these costs, along with electricity and water charges, steadily increased.

The Society had successfully negotiated a change of venue for its activities from the city centre (town gardens from 1851 and Market Square from 1854) to the drill hall site (from 1889) at its southern fringe to its current home (from 1902) at the opposite end of central Pietermaritzburg. It had suffered the consequences of both local and international military-political crises. These included the Langalibalele uprising (1873–1874), the Anglo-Zulu War (1879), the Anglo-Boer Wars (1881–1882 and 1899–1902), the Bhambatha uprising (1906), as well as the First (1914–1918) and Second (1939–1945) World Wars.

It had coped with sometimes disappointing attendance figures at its Royal and Garden shows, often due to unpredictable weather conditions or other public distractions. It also faced a declining membership and pool of voluntary assistants which, in part, was linked to changing regional demographic patterns. These were accompanied by differing interests, needs and expectations on the part of the general public concerning exhibitions and function venues.

Indeed, by 2019 the Society increasingly found itself wrestling with negative perceptions in some quarters as to what it actually represented and was trying to achieve. Its declared mission statement had nevertheless remained virtually unchanged since first being published in 1999, with the term 'not for gain' replacing 'non-profit making' in the first paragraph to conform with international norms.[3]

To its advantage the RAS had owned its 13.5-hectare Showgrounds since August 2006, after previously leasing them from the municipality. It enjoyed significant income via the Royal Mews Trust from leasing the extensive properties on its eastern boundary and from premises rented out within the grounds themselves, such as Chamber House which the PCB occupied.

The Society's sound financial condition, despite the prevailing unfavourable economic climate, was attributable not merely to gate takings and site rentals at the annual Royal Show but also to other events. From as early as 1896 and particularly from the 1950s, it derived additional income from the use made of its grounds and facilities for a wide variety of occasions which took place at other times of the year. These functions attracted exhibitors, participants and visitors from all over South Africa and abroad.

The RAS continued to maintain a sense of corporate responsibility even though it was a 'not for gain' organisation. It assisted appropriate, selected projects by, for example, waiving the hire charges for agricultural and quasi-agricultural conferences, events and training schemes. It also conducted its own training programme for 200 emerging farmers free of charge, set up a medium-term loan for a deserving future farmer to study abroad and mentored young learners, especially from the rural areas.

It provided home base facilities, donations and/or reduced the hire charges by up to 75% for certain worthy causes. These included the Deutsche Verein, the Education, Culture and Welfare Organisation, the Family Aid Trust, Gift of the Givers, Gujarati Vedic, Hospice, Lions International, PADCA, Rotary and Round Table, the Saibaba Foundation, the Salvation Army and the South African Guide Dogs Association. In 2018 and 2019, for example, the Society donated R18 000 and R11 000 to the SPCA from the proceeds of the Royal Show Gold Cup Charity Auctions.[4]

In October 1995 and again late in 2018 the RAS had streamlined its administrative structure to make it more efficient and cost-effective. Importantly, it had always enjoyed the services of a series of very capable executive and management committees. Included in their number was Dr Iona Stewart, the Society's first female president who was elected 159 years after its foundation and just over a century after it first admitted women as members.

Following the gradual relaxation of the government's admission regulations it was only in May 1978 that a blanket permit had allowed for full multiracialism with regard to both exhibitors and visitors. In 1986 the Society had also belatedly begun to admit 'all race groups' as members but, apart from the inclusion of local businessman Kay Makan and some City Council representatives, it had struggled to attract any other than white recruits onto its executive.[5]

It was also fortunate in being able to rely upon competent staff and a loyal if declining corps of volunteer helpers. Two decades into the 21st century this was as vital as ever in the ongoing climate of what its president Mike Moncur described as 'increased costs and persistent economic and political uncertainties'.

With national confidence still faltering it was also essential to maintain 'a policy of conservative prudence in all its dealings' while being aware of 'current risks and future possibilities'. Moncur contended that, until there was an economic upturn, the exhibition-based industry would remain 'tenuous', dependent as it was on 'limited client marketing budgets.' The prevailing watchwords 'caution and prudence' therefore remained applicable.[6]

Prospect

Under these circumstances the issues already raised in the SWOT analysis contained in the Society's 2016 strategic overview were still very relevant with perceived strengths, weaknesses, opportunities and threats largely unaltered.

Dramatic demographic changes in the region, white emigration and insufficient interest on the part of other ethnic groups seemed likely to maintain the longstanding decline in RAS membership, although there was a pleasing if unexpected increase in 2019–2020. This stood at 1 266, including 1 085 adult, 99 junior and 82 life members. The Society's interactive social media platforms continued to function effectively, but the volunteer pool was shrinking.

Several sections of the annual Royal Show, such as Sheep, Rabbits and Crafts and Home Industries, were now wrestling with the problem of replacing competent but aging judges with suitably qualified younger successors. Few, it seemed, found time for such hobbies although in 2019 the budgerigar, cage bird, poultry and rabbit categories of the Show were still strongly represented.

There was also the broader challenge of finding suitable young candidates, preferably drawn from both genders and all ethnic groups, to serve on the executive. The culture of club membership, voluntary service in shows and on committees seemed to have become incompatible with hectic modern lifestyles. It was suggested that financial incentives, if affordable, or other obvious benefits might improve the situation, but in some communities society membership and committee participation remained a culturally unfamiliar activity.

The annual Royal Show and the availability of the Showgrounds facilities for both public and private functions were certainly well-known across the social spectrum. Even so, in some quarters it was felt that the prospect of venturing into a society like the RAS, dominated as it was by white males and strong sectional identities, was decidedly daunting for women and members of other ethnic groups.

It was argued that the Society could have been less conservative, more proactive and more obviously welcoming in encouraging their involvement, thereby enlarging its membership and volunteer pool. In some communities pre-1994 socio-political barriers were not easily forgotten or broken down. These challenges were by no means unique to the RAS. Some golf clubs, for example, sought to increase their membership and income by constructing mountain bike tracks around their course perimeters.

The Society also faced the daunting question as to who would eventually succeed Terry Strachan as CEO. He was widely recognised as a highly successful general manager who had retrieved the RAS from the brink of disaster in the late 1990s and built it into a financially sound institution. As former vice-president Graham Atkinson observed in a personal tribute 'We as members, and the Royal as a Society, are greatly indebted to you for the sterling contribution you have made towards its progress and stability over the years since you joined the Society'.

However, by late 2019 there was no obvious candidate groomed to succeed Strachan in this crucial multi-tasking role. Careful introspection, strategic planning and innovative ideas as well as effective communication and marketing would clearly continue to be as important as ever. It would also still be essential not to compromise the Society's traditional commitment to promoting agriculture and agriculturally-related initiatives.[7]

By 2019 there were other significant new challenges. The national Association of Show Societies had virtually come to an end and in a debate that had been ongoing for years some questioned whether shows like the Royal still served a necessary purpose. They argued that it was 'a colonial importation' which for demographic and other reasons had, sadly, 'served its use by date'. Former presidents Derek Spencer and Garth Ellis were two of the strongest advocates of 'reshaping' the RAS for the future without necessarily retaining the Showgrounds.[8]

Indeed, the ongoing decline in the number of commercial farmers as well as the increasingly effective impact of video conferencing, online virtual business and computerisation now seemed to make such events irrelevant. The availability of internet information about pedigrees together with modern means of artificial insemination suggested that the physical presentation of prime livestock had become unnecessary. As Mike Moncur put it, there was no longer even any need 'for cows to come into contact with bulls'!

Conversely, this ignored the unquantifiable benefits of real face-to-face exhibiting and marketing as well as meet-and-greet contact and personal networking among farmers. Moreover, members of the public who were attracted to agricultural shows understandably expected to see farm livestock.

There was also the educative value, particularly for youngsters, of witnessing such farm activities as milking and sheep shearing in addition to learning about birds, rabbits and the manufacture of honey. Marlize du Preez and others argued that the Royal Show had survived for so long because it had managed to remain relevant and in these and other innovative ways could continue to do so.

However, as in other parts of the world, farmers were facing mounting labour and transport costs associated with bringing animals to shows held in towns as well as keeping them there for long periods. In addition, there were security issues arising from their extended absences from the farms. There was also increasing competition from weekend markets and some exhibitors were switching their allegiance to them. Commercial exhibitors were similarly confronted with rising costs and argued that the Royal Show's ten-day duration was one of its biggest drawbacks. Both agricultural and commercial exhibitors alleged that local B and Bs inflated their prices during show times.

Nevertheless, early in 2019 Terry Strachan expressed confidence that while some potential exhibitors understandably baulked at the expense of participation and needed subsidisation or sponsorship, the farming community within and beyond the province still regarded the Royal as South Africa's 'premier agricultural show'. Moreover, it was still 'probably the largest mixed agricultural show in the country catering for competitive livestock and equipment', second only in size to the Free State's enormous Nampo Show at Bothaville which was 'almost entirely implement and machinery orientated'.[9]

Among its other challenges the RAS was confronted by the mounting pressures of urban spread. Its Showgrounds were situated in an area that had increasingly become a residential suburb to the north-west and a commercial zone to the south-east. In December 2018 it was reported in the local press that the Society's entire property might be sold to an as yet unidentified developer, possibly to establish a specialist training hospital and conference facilities that would further enhance the city's standing as a prominent medical centre.

As early as the 1940s there had been talk of moving the Showgrounds to Mkondeni where the overhead costs of a huge complex might be shared with sporting bodies such as polo and the Pietermaritzburg Turf Club. The prospect of losing the Society's individuality and already established facilities in such an amalgamation, coupled with the expense involved, had eventually led to the proposal being rejected in 1965. Nevertheless, well into the 1970s there were those who continued to favour the move, to no avail.

In October 1997 cattle breeder John Fyvie had again suggested, in view of the frequent flooding of the grounds, that it might be time to relocate. He was informed that 'this option had been explored and for various reasons, was not acceptable'. In 2003 the RAS had been involved in discussions concerning yet another proposed major development, this time at Ashburton, adjacent to the N3 motorway, in which it was envisaged that it would be apportioned a substantial share of the land.

By contrast, in the wake of earlier forward planning devised in the 1970s, a 2006–2007 master plan had visualised the development of the Society's existing Showgrounds into four components. These, it was hoped, might integrate them more effectively into the commercial requirements of the city while generating much-needed income and still promoting the original agricultural core of its activities as specified in its constitution and mission statement. In April 2015 the executive had tasked a sub-committee 'to consider the Society's future, especially in respect of fulfilling the organisation's constitutional *raison d'être*'. What followed was the 2016 strategic overview.

In the same year Rob Haswell had proposed that, with the KZN Rugby Union contemplating a move from the Woodburn Stadium to make way for commercial expansion, the appropriate development of the Showgrounds Main Arena might provide an ideal new home for that sport in the Midlands. It was envisaged that this could also accommodate the professional Sharks rugby team, as well as serving the needs of T20 cricket tournaments and major concerts. For some time, nothing more was heard of this proposal and eventual discussions with rugby union officials did not take the matter any further.

By 2018–2019 there were rumours that plans were now definitely in place to sell the Showgrounds and move to new premises further inland to the vicinity of Howick; or, more likely, south of Pietermaritzburg and nearer to the greater Durban region. This was still the most promising local catchment area to attract public interest in the Society's various activities.

Terry Strachan did concede that, faced with rapidly changing circumstances, the RAS might well accept a serious offer to buy its property but, in an effort to curtail misinformed speculation, in 2019 members were assured that no such plan to sell the Showgrounds had been formulated. He informed the *Witness* that there had been 'nibbles, not definitive expressions of interest' in the past but that no firm proposal to purchase was as yet being considered.

RAS president Mike Moncur believed that at some stage it might well be necessary to move in view of the advance of commerce but that currently it was still a case of 'business as usual'. If the Showgrounds were indeed sold, it would be 'a sad day' but they would have to be disposed of en bloc and not piecemeal. Any move would have to be to an appropriate site towards Durban with adequate access and parking space for exhibitors and the public. Such a development would probably also suit other Showgrounds activities such as the Garden Show.[10]

The withdrawal of agricultural events to peri-urban sites was already an established trend abroad in the face of parking and traffic issues, rising municipal tariffs and rates as well as the expensive logistics of moving large numbers of animals in and out of town. For example, in 1998 Australia's biggest single annual event, the Sydney Royal Easter Show, gave way to urban expansion and, after 116 years, moved with great success from Moore Park to the Sydney Olympic Park area.

Nearer to home, Johannesburg's Rand Show/Rand Easter Show was initially held at the Old Wanderer's Cricket Ground in 1894 before being shifted to Milner Park the following year until 1984. It had then moved again when the Rand Show brand was sold to the Johannesburg Expo Centre with the intention of being incorporated into the Easter Festival at Nasrec. The move from Milner Park also made it possible to develop what became the University of the Witwatersrand's West Campus.

The subsequent 2009 Johannesburg Easter Festival was reportedly not as successful as had been hoped. It lacked the familiar atmosphere, attracted disappointing attendance and there was some dissatisfaction on the part of exhibitors. Proposals to change the name of Pietermaritzburg's Royal Show in 2010 were promptly dropped.[11] In 2013–2014 the Pretoria Show ended with most of its exhibitors transferring to Thabazimbi.

The 2016 RAS strategic overview had already indicated that, as its own situation altered and with its 'primary activity [the Royal Show] being increasingly questioned', changes to the business model the Society currently operated might become unavoidable. It had concluded that the status quo could probably only survive in the medium term by relying primarily on financial support from the Royal Mews Trust, but that the latter only had 'a finite revenue generating capacity'.

Alternatively, the status quo might be sustained much longer by making more Showgrounds land available to the Trust for further commercial development and greater income generation. Implementation of this option would probably be at the cost of holding smaller and shorter shows which might not be readily acceptable to exhibitors and other stakeholders.

The strategic overview had indicated that a third possibility might be to retain the Royal Mews Trust development zone and sell the Showgrounds – as was suggested in the media little more than two years later. It was speculated that such an event might raise an estimated R150 million which could be invested to generate income supplementary to that provided by the Royal Mews Trust, amounting to R3 million a year by 2016. The RAS would then have the means to fund worthy projects related to agriculture, such as the

Eston and Underberg shows, thereby returning to its 'initial core values' while continuing to preserve and grow its financial reserves.[12]

Terry Strachan surmised that, if the latter option was adopted, the Society could follow the lead of the Cape of Good Hope Agricultural Society (Agriexpo) which had sold its Woodstock property 20 years previously to give way to a casino, retaining only a small piece of real estate from which to operate. In doing so it still had the financial means to support agricultural events such as the annual Cheese Expo and South Africa's biggest dairy show.

Garth Ellis contended that the future of the RAS was under serious threat, that change was essential although it could not happen overnight, and that there was actually no longer any need for the Society to own land. Similarly, Andrew Line envisaged no more than 'a 50m^2 office with ten people working from there to assist agricultural shows around the province'.

Conversely, it was argued that such a course of action would probably not work in KwaZulu-Natal because the situation was 'totally different to the Western Cape' which was better suited to 'smaller shows', such as the dozen or so poultry exhibitions held there each year. In addition to the Royal Show, other major local events such as the Garden Show would then either come to an end or have to find another venue.

On the other hand, annual Pietermaritzburg attractions such as Cars in the Park and Art in the Park had seemingly moved with success, the former from Alexandra Park near the centre of town to peri-urban Ashburton and the latter to the Botanical Gardens in suburban Prestbury.

In 2018 Manco discussed the possibility that the RAS might take the same route as Scotland's Royal Highland and Agricultural Society and the Royal Easter Show in New Zealand. These, as well as several in England, Australia and Canada, involved minimal permanent infrastructure and therefore far less maintenance, being held in modern tubular-framed marquees with portable stalls and stables.

Such an option would involve finding a suitable rural base, perhaps 15 or 20 hectares in the Natal Midlands with convenient access to Pietermaritzburg and Durban, on which to maintain its fundamental commitment to agriculture but continue to hold other events.

Not everybody considered it a viable option but Mike Moncur envisaged that, in such an event, there should be sufficient land not only for livestock and multi-disciplinary equestrian events but also to hold Nampo-like demonstrations of equipment and implements which were not currently possible in the existing Showgrounds.

Sentimental attachment to the Society's well-established home, with its wide variety of facilities that could cater for virtually every occasion, was nevertheless still very strong. Tim Nixon pointed out that there would be challenges in replicating this and especially in recreating a similar atmosphere. Such a model would have to be carefully packaged and re-marketed as something new rather than as the same show at a different venue.

It could nevertheless continue to provide a base for the various feather and fur clubs that still regarded the Showgrounds as their home. There would also need to be sufficient space for livestock exhibitions for which there was still a significant demand among farmers, especially if held in a more accessible peri-urban setting and for a more limited period of time.[13]

Whether such an arrangement would indeed continue to satisfy the requirements of the numerous other annual activities taking place on the Society's premises, not least the Garden Show, or prompt their departure elsewhere, remained uncertain.

Moreover, as had been stressed at the 2008 RASC conference in New Zealand, if moving to a new site became necessary as had already happened elsewhere, it was essential to 'have all the funding in place in advance' as 'having only a portion will in most cases be a catalyst for disaster'; further, that it was also vital 'to have a political friend and champion, who will stay with you and fight your cause'.

The future of the RAS clearly depended heavily upon financial considerations and any sale of the existing Showgrounds would almost certainly have to await a substantial economic upswing before a commercial buyer could be found. Alternatively, it might provide a suitable site for the proposed new provincial parliament and related office complex, previously envisaged for the nearby Town Hill Hospital grounds, assuming that sufficient public funds were available for that purpose.

In mid-2019 Kay Makan estimated that the Society's property was one of the most expensive pieces of real estate in the city being worth at least R300 million. He considered it unfortunate that it had not been sold a few years previously and doubted if any interested party now had the means to buy and re-develop it in its entirety.

Garth Ellis agreed that the optimum time to sell the Showgrounds for a good price may already have passed, not least in view of the municipality's apparent incapacity to provide the infrastructure necessary to support any re-development. Other disturbing possibilities were that the RAS might be pressurised to part with its property by steep increases in rates and the cost of municipal services, or that the State might simply resort to expropriation, with or without compensation!

In any event, as Nick Proome of Elphick Proome Architects pointed out, re-development of the Showgrounds premises would still have to await environmental impact studies and land rezoning applications which could take two years to process.

Twelve years previously Proome's firm had been involved in re-developing the old fairground at the northern extremity of the Society's property into the Bank Park which Standard Bank and Absa now occupied. He reportedly envisaged the possibility of an integrated mixed 'work, live, play' (that is, a business, residential and recreational) zone that would ensure 24-hour use of the old Showgrounds site.[14]

Any such changes would be subject to appropriate urban and architectural design controls which gave due recognition to the heritage of the city and its capital status. Apart from the municipality's capacity to provide and maintain the necessary infrastructure, the possible redevelopment of the Showgrounds also prompted speculation whether this would attract further investment to Pietermaritzburg and create much-needed job opportunities.

For some the debate evoked recollections of the Royal Show that were far from happy. They remembered a time when it 'was only open to non-whites on the last day' and some exhibitors had 'already packed up to leave'.[15] Others were fortunate enough to enjoy childhood memories of exciting visits to the Royal Show and its funfair, as well as of more recent Garden Shows, banquets and similar functions.

The discussion also highlighted the longstanding role the Society had played in the life of Pietermaritzburg. A Matrix survey conducted in 2002 had already affirmed that it attracted more visitors to the city, as well as their spend money, than any other activity had ever done and indirectly had created 2 000 permanent jobs in the uMgungundlovu region. In Terry Strachan's estimation by 2019 the RAS and its shows had become one of the largest, if not the biggest, contributor to the local economy with the multiplier effect of its presence injecting R400 million a year into the region.

Indeed, it had played this role for so long that it had largely come to be taken for granted, most notably by the civic and provincial authorities who, it was felt, made no effort to market the Society's activities. This was in striking contrast to events held at Durban's ICC which enjoyed active support in recognition of their importance to the city's image and coffers, not to mention the generous public subsidies that agricultural organisations elsewhere received as a matter of course.

By 2019, among the numerous other events which the RAS hosted each year, the Royal Show was still the most important and retained its status as 'the largest mixed exhibition incorporating a fully fledged agricultural component on the continent'. With the 45th Garden Show on the horizon the following year and its own 170th anniversary in 2021, it could reasonably be expected that the Society would continue to make a significant contribution to the city and to the country.[16]

To that end there were promising developments early in 2020 with preparations for the next Royal Show already well in hand. The River Stage was to be relocated to a part of the Clock Tower bank, thereby effecting a R70 000 saving. A new entrance to the Olympia Hall had already been completed with a reconfigured demo kitchen also envisaged there. In addition, Iron Horse Productions were commissioned to develop a Theatre of Meat at the centre of the hall which was intended to involve at least fifteen meat-related exhibits.

After intense negotiations a long-term protocol was concluded by the six largest shows in South Africa, including the Royal. The finances of the RAS were still sound and Terry Strachan was in discussions with the municipality about altering the commercial description of the Showgrounds in order to reduce the rates burden. A tight rein was being maintained on expenditure and in January a meeting of former presidents was held to update them on the current situation.

At the end of February 2020, the combined financial results of the Society and the Trust reflected a surplus for the first time in several months. In early March the Showgrounds once again provided a venue for the opening of the provincial legislature, the state of the province address and the presentation of the budget. Among other smaller functions there was also a Department of Social Development Wellness Day and the Witness Careers Expo. These events further improved the financial situation and there was an encouraging increase in bookings for use of the premises.[17]

Unfortunately, 2020 proved to be a year of unexpected crises. A foot and mouth outbreak in Limpopo necessitated Terry Strachan's attendance at a meeting in Pretoria with representatives of Agri-Expo, Alpha, the Bloem Show and Nampo. The expert information provided there by veterinarians was subsequently shared with all potential Royal Show livestock exhibitors. A positive outcome was the formulation of the national biosecurity protocol.

The Society's fortunes deteriorated dramatically with the arrival in South Africa of the Covid-19 pandemic. It was a crisis at least as potentially devastating as any other which it had faced in its long history as the exhibition and conference industry evaporated overnight.

Following the declaration on 26 March of a national lockdown, by month's end two major events scheduled to be held on the premises had already been cancelled: the UKZN graduation ceremonies and a Department of Agriculture summit. More were to follow, including a DUT graduation and numerous business and social functions.

By the end of April, it had been decided to cancel the Royal Show scheduled for July that year in view of current scientific estimates that the incidence of virus cases would only reach a peak in September or even October. This uncertainty, coupled with the government's extended lockdown strategy to reduce its impact, reluctantly obliged the RAS not to reschedule the Show for later in the year. It was decided that exhibitors, and particularly farmers, could not be expected to prepare for an event that might not take place at all.

Moreover, there was concern about the attendance of large numbers of people, including school groups, at a time when this might still not be recommended or even permitted in terms of the lockdown regulations. At least one section, the Natal Rabbit Club, arranged its own Royal Show by means of a virtual exhibition with six photographs of each of the 184 entries being submitted for judging, including one from Botswana and another from Scotland. Despite its shortcomings, this was an innovative means of maintaining interest in the hobby during a difficult year.

It was only the second occasion on which the Royal Show had been cancelled due to global circumstances, the first being World War II. The obligatory refund of stand rentals to exhibitors who were not willing to roll these over to 2021 amounted to approximately R700 000. Moreover, there was to be no significant reduction in fixed costs and service providers who had already met their requirements had to be paid.

The cancellation of what was its primary source of income was a major financial setback not only for the RAS but for the Pietermaritzburg region, which stood to lose at least R250 million in revenue. The Society continued as best it could to protect its cash reserves but the cancellation of numerous other significant income-generating events followed, including that year's Witness Garden Show.

Some of these, like the budget sittings of the provincial legislature, followed a now increasingly fashionable virtual format which, it was feared, could become commonplace in the future to the disadvantage of the RAS. Difficult decisions had to be taken in the interests of keeping jobs and grounds maintenance. An extra R60 000 had to be found for an untimely but unavoidable upgrade of the Society's computer server and another R35 000 for the replacement of the Royal Food Court freezer.

In collaboration with the provincial Department of Health, it was decided to convert Halls 6, 7, 8 and the Olympia Hall, all to the north of the Dorpspruit, into a Covid-19 quarantine step facility where patients from high-density environments who were awaiting the results of their tests could be accommodated before being discharged if deemed negative; or immediately relocated to a provincial hospital if not.

The purpose, as in other parts of the country, was to provide support in the event of conventional hospitals being overwhelmed by a flood of virus and suspected virus victims. Similar facilities were established at the Clairwood, Wentworth, Richmond, Dundee and Niemeyer hospitals as well as the Durban Exhibition Centre.

It was intended that no known Covid-19 patient would be retained in the Showgrounds. Further, that the halls used for quarantine purposes would have entirely separate entrances to the premises and would be completely isolated by fencing and security services from those parts used by other organisations.

These included Chamber House which was occupied by the Pietermaritzburg and Midlands Chamber of Business. Sadly, following a prolonged disagreement concerning ownership and the payment of rental, the latter gave notice of its intention to vacate the premises on 30 June after a 36-year presence in the Showgrounds.

Early in May the completed 254 bed cubicles met with the approval of local municipal officials and subsequently that of President Cyril Ramaphosa, who visited the halls accompanied by provincial premier Sihle Zikalala. This new field hospital was formally opened on 14 June, complete with its own resuscitation bay as well as Covid-19 resistant antimicrobial and sporicidal disposable curtains.

The intention was that it would serve the needs of patients in the uMgungundlovu, Harry Gwala and uThukela districts, possibly for as long as five months. Strachan estimated that this agreement would net the RAS approximately R500 000 per month or R2 million over a four-month period at a time when, apart from rentals, it enjoyed no other sources of income.

This made it possible to cover fixed costs and protect cash flow during the last two months of the financial year ending on 30 June 2020. Unfortunately, it still reflected a net loss of R1 268 553, the largest in the Society's history, although a strong balance sheet and the reserves accumulated during several years of prudent financial control made it possible to avoid a complete catastrophe. Moreover, the Royal Mews Trust's surplus of R2 469 586 generated that year reflected a pleasing increase of R139 986 over the previous year.

By the end of September 2020, the Showgrounds halls were being decommissioned along with other field hospitals following a significant decline in the Covid-19 infection rate. By then the KwaZulu-Natal Department of Health was confident that it now had sufficient capacity of its own to cope with any possible second wave of infections. Within a month the halls were again available for conventional purposes after being professionally sanitised and decontaminated.[18]

To add to the gloom an unhappy year witnessed the deaths of more RAS stalwarts and associates. Late in 2019 these included Bobby Hex, friend, advisor and Exco member, Neil Raw, for many years a bird hall exhibitor and supplier of livestock bedding, and Andy Foulis, Barbara Shaw's partner. They were followed in 2020 by honorary life president Dr Max Taylor, Syd Whelan, husband of Henrietta Whelan, and grounds superintendent Roy Motilal, the latter after 36 years of service .

The RAS also mourned the loss of the much-loved Ken Easthorpe who had been associated with the RAS for more than 60 years, former president and honorary life president Ron Glaister and Joe Spencer, a senior *Farmers' Weekly* journalist and loyal friend of the Royal Show. Early in 2021 Prince Phillip, HRH the Duke of Edinburgh and Lord Samuel Vestey died. Both had given long service to the Royal Agricultural Societies of the Commonwealth and had always showed great interest in the activities of the RAS of Natal.

Other losses included Garth Carpenter who had overseen the Royal Show Snake Park for many years, small stock exhibitor Avison Carlisle, chairman of the Pigeon Club Jan Oelofse, function co-ordinator of the Pietermaritzburg and Midlands Chamber of Business Lorna Jones, honorary life member and longstanding senior livestock steward Brian van der Bank and Jeremy Jonsson, who had been actively involved in the Show's Dairy Section for more than 50 years.

On a more cheerful note, honorary life membership was conferred on Robert Hoekstra, founder and CEO of Stromberg, in recognition of his 60-year association with the Show and on Neville Thomas who had served on the Society's executive committee for ten years. So, too, had vice president Tim Nixon who became honorary life vice-president.[19]

In addition to the Covid-19 pandemic, other significant developments took place during the course of 2020 and 2021 that would determine the future of the Society and its Showgrounds. In a mid-July 2020 circular letter CEO Terry Strachan reminded members that during the previous few years they had been kept informed in annual reports, at annual general meetings and through the media about the options being considered with regard to strategic planning for the future.

He stressed that it had become increasingly clear in the light of 'current realities' and 'persistent deficits' that the RAS now needed to develop a new business model. Strachan reminded members that this could take the form of the Cape Show (now known as Agriexpo), which had relinquished its Woodstock Property twenty years previously but was now 'more meaningfully involved in agribusiness than at any time in their history'. Alternatively, it could follow the example of the Royal Highland and Royal New Zealand shows. This would involve acquiring 15 to 20 hectares of land, either freehold or leasehold and preferably close to town but with minimal built infrastructure, on which to hold agricultural and multi-disciplinary equestrian events.

Strachan informed members that after being approached on several occasions during the previous decade by developers with tentative offers, he was able to announce that, despite the country's ongoing economic recession, the Society had recently received no less than 'three meaningful proposals, each of which with the input of professional advisors are currently being considered'. Consequently, more so than at any other recent time, it was possible that a sale agreement might soon be negotiated.

He gave the assurance that, subject to the then prevailing health regulations, the 2021 and probably the 2022 Royal Shows would still be held at the current address and that any sale proceeds that might be generated in the interim would be securely held in trust quite separately from day-to-day RAS operations. Further, that the survival of the Society was 'non-negotiable' and that it would continue to be 'unequivocally committed to remaining a conduit of support for agricultural endeavours'.[20]

What had been a particularly grim year ended with a cheerful Carnival of Lights at the Showgrounds including a funfair, craft market, outdoor movies, food court and bar. The easing of Covid-19 restrictions had already made it possible to begin hosting smaller functions. Moreover, preparations were well underway for the 2021 Royal Show which was to coincide with the Society's 170th anniversary.

Ominously, in the face of a rising second wave of the pandemic, the Carnival of Lights ended on 18 December instead of 30 Decemberr and it remained to be seen when any events would again be possible. As 2021 approached there was still reason to believe that the RAS was on the brink of a new dawn, as portrayed by Terry Strachan's photograph on the cover of the 2020 *Annual Report* of a September sunrise breaking over the Main Arena.

Unfortunately, that change of fortune was, at best, to be delayed by yet another disastrous year of pandemic and lockdown. Before the end of December 2020, it was announced that in view of the new spike in infections the 254-bed Showgrounds field hospital was to be recommissioned as a precautionary measure. By 19 January when it was re-opened the province's public and private health facilities were under severe strain with 5 000 new cases being recorded each week.

Two days later RAS members were advised that, 'owing to uncertainties relating to COVID-19', the Royal Show was regrettably to be cancelled for the second year in a row. Membership subscriptions were again to be carried over and it was likely that the 2021 Garden Show would also not take place. The financial and other implications clearly extended far beyond membership numbers and subscription income.[21]

By May 2021, with a third wave of the pandemic imminent, the Showgrounds field hospital had assumed a further dimension when it became an important venue for the government's anti-virus vaccination programme as it at last gathered momentum. Health care workers and then members of the public over the age of 60 were the focus of its first two phases with the intention that the rest of the population would eventually follow.

The financial benefit derived from providing a site for medical purposes offered substantial compensation for the huge loss of income resulting from the minimal use of Showgrounds facilities for other events during the pandemic. One significant exception was the large gathering to select a new Anglican Bishop of KwaZulu-Natal held in May 2021 during the brief lockdown relaxation prior to the Covid-19 third wave.

The Oxford Old English Game Fowl Club still managed to host its annual show while the Natal and Coast Poultry Club held two successful events, which indicated that there was still a willingness to participate in shows. Selected members of the Rabbit Section took part in virtual shows abroad and in the Gauteng Breeders championship.

There was little or no activity in other sections but, on behalf of the RAS, its president Mike Moncur fervently hoped that 'shows, exhibitions and sporting events' would soon 'return to normality'. To that end plans were afoot for the 2022 Royal Show when the Society's 170th anniversary late in 2021 would be celebrated.

In the interim the RAS understandably did not attract any sponsorship during the 2021 financial year. Due largely to the income derived from the field hospital and vaccination centre it nevertheless enjoyed a pre-taxation surplus of R1 827 799. This amounted to one of its better financial outcomes during the previous decade. The lease agreement with the provincial Department of Health was not due to expire until January 2022 and there were sufficient funds to buy a generator as a backup for Halls 6, 7 and 8 while the refurbishment of the Umgeni Walkway was also completed.

The impact of the ongoing pandemic on the events industry had an understandable effect on staff morale and, due to the lack of work in their departments, Lauren Fisher, Lizette Storey and Irene Peters had to be placed on extended furlough with the former two finding alternative employment. Happily, the Society's BEE rating improved from level 5 to level 4 and Messrs Cele, Ngidi and Ndela were presented with awards for 30 or more years of service.[22]

In May 2021 the ongoing uncertainty about whether the Society would soon assume an entirely new format or maintain its familiar presence was to some extent clarified. In a notice to all members, Terry Strachan announced that on 25 November 2020 the RAS had accepted an offer from Vu-Tact Trade and Invest (Pty) Ltd to purchase its Showgrounds. The intention was that under new ownership they would eventually become the site of a 'mixed-use' development involving retail, office and residential dimensions.

This agreement excluded the Royal Mews Trust properties. These now included the portion occupied by Hereford Financial Services following the provision of an electrical supply independent of the Showgrounds, which facilitated the transfer to the Trust and the consolidation of the Society's remaining holdings.

Westville architect Nick Proome, who had earlier been involved in converting the old fairground into a bank park, was a member of the purchasing consortium. He affirmed that, as far as future growth was concerned, the Showgrounds were 'one of the most valuable sites in Pietermaritzburg'. The intention was to develop it in an aesthetically pleasing manner with buildings conforming to a cohesive theme.

Melanie Veness, CEO of the Pietermaritzburg and Midlands Chamber of Business, considered it inevitable that such a valuable property would sooner or later be redeveloped and that this would boost the local economy as well as offer the prospect of more employment opportunities. Msunduzi mayor

Mzimkhulu Thebolla welcomed these possible outcomes, but was concerned about the loss of the large conference facilities which the Showgrounds offered and the possible adverse impact on tourism.

Terry Strachan and Mike Moncur agreed that it had become impossible to hold an agricultural show involving 5 000 head of livestock and other heavy traffic in what was now part of the city's central business district. Strachan again gave the assurance that 'the future of the RAS and the Royal Show are not in doubt'. The Society, with all movables, would find a more suitable home, hopefully within the next year and possibly in the Howick/Lions River/ Midlands region, but preferably somewhere in the arc to the south of the city from Umlaas Road to Baynesfield.

While there was understandably a great deal of nostalgia attached to the old Showgrounds. many like Mooi River beef farmer and executive committee member Angus Williamson considered these developments opportune and would facilitate the expansion of the agricultural dimension of the RAS, not least with regard to livestock.

It was expected that the move would not take place for at least two or three years as the sale agreement involved certain suspensive conditions, including a crucial environmental impact assessment and other feasibility considerations. Consequently the 2022 and 2023 Royal Shows were still expected to be held in Pietermaritzburg, virus pandemic permitting.

It was envisaged that at a new fit-for-purpose facility the Royal Show would in future possibly run for only six days, from a Friday to the following Wednesday. It would indeed comprise a substantial livestock show, as well as equestrian and agricultural equipment components, a 'moderately sized commercial section' and a weekend music festival involving 'three or four genres'. In combination this 'would constitute the main income stream' for the RAS. There would be no funfair, but the Apiarian, Bird, Crafts, Home Industries and Rabbit sections would still be involved.

The Society's anticipated return to its 'agricultural roots' on new premises in a not too-distant, peri-urban setting promised in large measure to maintain its longstanding traditions. It might then still be the case, as sometime vice-president William Dreboldt liked to put it, that 'at Show time all roads lead to Pietermaritzburg!'[23]

Hopefully, in whatever form, the RAS will in the future continue to be an invaluable asset to the city, the region and indeed the whole country.

ENDNOTES

1 *AR* 2019: 4, 6. *Witness*, 25 June 2019: 1.

2 Terry Strachan, interview, 12 March 2019.

3 *Witness*, 26 July 2018: 10. *AR* 1999, 2018: inside front covers. RASM, Management, 13 March 2007: 2.

4 RAS *Prospectus 2012–2015*: 16; *AR* 2019: 23.

5 Shute, 'Royal Show': 9.

6 *AR* 2017: 24, 2018: 22.

7 RAS, Office of the CEO, 'Strategic overview': 9–13. Di Fitzsimons, interview, 4 June 2019. Ron McDonald, interview, 12 June 2019. Mike Moncur, interview, 26 June 2019. Kay Makan, interview, 2 June 2019. Tim Nixon, interview, 1 August 2019. *AR* 2020: 9, 16; Janice Will, e-mail, 20 May 2021 attaching Graham Atkinson, e-mail to T.D. Strachan, 16 May 2021.

8 Derek Spencer, interview, 29 May 2019. Garth Ellis, interview, 13 September 2019.

9 Terry Strachan, interview, 12 March 2019. Derek Spencer, interview, 29 May 2019. Di Fitzsimons, interview, 4 June 2019. John Fowler, interview, 10 June 2019. Mike Moncur, interview, 26 June 2019. Tim Nixon, interview, 1 August 2019. Marlize du Preez, interview, 5 August 2019. *Witness*, 7 June 2019: 9.

10 *Witness*, 3 December 2018: 1. Terry Strachan, interview, 6 March 2019. Mike Moncur, interview, 26 June 2019. Bert Cornell, interview, 18 June 2019. Marlize du Preez, interview, 5 August 2019. RASM, Cattle, 6 October 1997: 3; Management, 15 July 2003: 3, 19 January 2016: 3, 28 June 2016: 2 and 19 October 2016: 2; Rob Haswell, 'The case for a multi-purpose arena in Pietermaritzburg', 16 January 2016: 1–3; Executive, 22 April 2015: 5. Shute, 'Royal Show': 46. Gordon, *Natal's Royal Show*: 68–69, 89–90, 101.

11 *https://en.wikipedia.org/wiki/Rand_Show*. RASM, Executive, 27 January 2009: 5 and 28 April 2009: 4.

12 RASM, CEO's report, 18 March 2014: 3. RAS, Office of the CEO, 'Strategic overview': 14–26.

13 Mike Moncur, interview, 26 June 2019. Andrew Line, interview, 29 July 2019. Tim Nixon, interview, 1 August 2019. Garth Ellis, interview, 13 September 2019. *AR* 2019: 24–25.

14 *Witness*, 3 December 2018: 1. Iona Stewart, interview, 12 February 2019. Terry Strachan, interview, 12 March 2019. John Fowler, interview, 10 June 2019. Ron McDonald, interview, 12 June 2019. Kay Makan, interview, 2 July 2019. Garth Ellis, interview, 13 September 2019. RASM, General Manager's Notes and Observations on RASC Conference November 2008, 27 January 2009: 1; Management, 17 April 2018: 2.

15 *Witness*, 12 December 2018: 10.

16 *Witness*, 26 May 2018: 12, 2 June 2018: 10, 4 December 2018: 6. *Weekend Witness*, 8 December 2018: 2. Gordon, *Natal's Royal Show:* 73, 118. Terry Strachan, interview, 12 March 2019; RASM, T.D. Strachan Report on Midi Strategic City Summit, 29 October 2009: 1–3.

17 RASM, CEO's Report, 22 January 2020: 1–3 and 25 March 2020: 1–2; Executive, 22 January 2020: 2, 4, 6. *AR* 2020: 15–16.

18 *Witness*, 29 April 2020: 1, 1 May 2020: 2, 5 May 2020: 3, 15 June 2020: 2, 26 September 2020: 5. RASM, CEO's Report, 22 January 2020: 2–3, 25 March 2020: 2, 22 May: 2–3 and 24 June 2020: 2; Executive, 24 June 2020: 4–5. *AR* 2020: 4–6, 8, 16, 17, 28–31. Melanie Veness, interview, 11 December 2020.

19 RASM, Executive, 22 January 2020: 1; CEO's Report 25 March 2020: 2 and 24 June 2020: 2. T.D. Strachan, e-mail, 7 October 2020. *AR* 2020: 10, 17, 2021: 2, 8–10.

20 T.D. Strachan to all RAS Members 'Re: future strategic planning', 14 July 2020. RASM, Executive, 24 June 2020: 1–3.

21 *AR* 2020: cover, 18–19 and pamphlet insert. T.D. Strachan, e-mail, 17 December 2020. T.D. Strachan to all RAS Members '2021 Royal Show: cancellation and membership matters', 21 January 2021. *Witness*, 31 December 2020: 1, 19 January 2021: 2.

22 *AR* 2021: 5–12.

23 T.D. Strachan, 'Note to members', 14 May 2021. *Witness*, 1 May 2021: 3, 15 May 2021: 5. Cynthia Dreboldt, interview, 5 March 2018. *AR* 2021: 5–10.

APPENDIX: THE RAS SINCE 1984

All dates of office listed are as given in the *Annual Report* of 30 June each year. * denotes member of the executive/management committee prior to 2018.

A. Patron

1984–	The State President

B. Vice-Patrons

1984–1994	His Honour the Administrator of Natal
1994–	The KwaZulu-Natal Premier

C. Presidents of the Society

1981–1987	R. McDonald*
1987–1991	J.M. Fowler*
1991–1995	R.J. Glaister*
1995–1998	H.D. Spencer
1998–2003	D.A. Wing*
2004–2006	A.J. Line*
2007–2009	G.I. Ellis*
2010–2012	Dr I. Stewart*
2013–	M. Moncur*

D. Vice-Presidents

1984–1987	J.M. Fowler* and J.S. Snaith*
1988–1990	R.J. Glaister* and R.B. Lobban*
1991–1995	H.D. Spencer*
1991–1993	Dr M.J.O. Taylor*
1994–1998	G.W. Poole*
1997–1998	J.F. Plummer*
1998–1999	W.C. Dreboldt*
1998–2008	G.D.J. Atkinson*
2000–2001	R.G. Brown*
2001	P. Breytenbach*
2002	E.A. Krause*
2003	A.J. Line*
2004–2006	G.I. Ellis*
2007–2009	Dr I. Stewart*
2008–2012	M. Moncur*
2010–	K. Makan*
2013–	T. Nixon*

E. Honorary Life Presidents

1984–1995	D.H. White-Cooper*
1988–	R. McDonald*
1991–	J.M. Fowler*
1996–2020	R.J. Glaister*
1999–	H.D. Spencer*
2004–2013	D.A. Wing*

2007–	A. J. Line*
2010–	G.I. Ellis*
2013–	Dr I. Stewart*

F. Honorary Life Vice-Presidents

1980–1984	Miss J. Fraser*
1984–1988	C.A. Filday
1984–1986	C.P.W. Francis
1984–2001	R.W. Hardingham
1984–1992	G.A. McIntosh*
1984–1993	I. Meyer
1984–1998	I.B. McFie
1984–1986	J.A.M. Shepherd
1984–1998	H.R. von Klemperer*
1988–1993	J.S. Snaith*
1988–1990	N.R. Pinnell*
1990–2001	R.G. Brown*
1991–2012	G.W. Poole*
1993–2015	R.B. Lobban
1994–2020	Dr M.J.O. Taylor
1995–	K.R. Howes*
2010–	G.D.J. Atkinson
2021–	T. Nixon

G. Honorary Presidents

1984–	Her/His Worship the Mayor of Pietermaritzburg/Msunduzi
1989–	The Judge-President of Natal/KwaZulu-Natal
1996–	KwaZulu-Natal Minister of Agriculture

H. Honorary Vice-Presidents

1984–	Her/His Worship the Metropolitan Mayor of Durban/eThekwini
1984–2002	The Chairman of the Afrikaanse Sakekamer, Pietermaritzburg
1984–1992	The President of the Durban Metropolitan Chamber of Commerce
1984–1996	The President of the Natal Agricultural Union
1984–1992	The President of the Natal Chamber of Commerce
1984–1992	The President of the Pietermaritzburg Chamber of Commerce
1984–1992	The President of the Pietermaritzburg Chamber of Industries
1985–1992	The President of the Natal Chamber of Industries
1986–	The Chairman of the South African Sugar Association
1993–2002	The President of the Pietermaritzburg Chamber of Commerce and Industry
1993–2001	The President of the Durban Regional Chamber of Business
1997–	The President of the KwaZulu-Natal Agricultural Union (Kwanalu)
1998–2001	The Chairman of the iNdlovu Regional Council
2002–	The President of the Durban Chamber of Commerce and Industry
2002–	The Chairman/Mayor of the Umgungundlovu Regional District Municipality
2003–	The President of the Pietermaritzburg Chamber of Business
2003–	The President of the Women's Institutes of KwaZulu-Natal
2003–	The Executive Director of Forestry South Africa

I. Trustees

1984–1987	R. McDonald
1984–1993	I. Meyer
1988–1990	J.M. Fowler
1991–1995	R.J. Glaister
1994–2016	R. McDonald
1996–1998	H.D. Spencer
1999–2005	J.M. Fowler
2006–2016	R.J. Glaister*

N.B. The office of trustee was dispensed with in terms of amendments to the RAS constitution in 2016.

J. General Committee Members

* denotes sometime member of the executive committee

1984–1993 T.W. Bailey; 1984–1989 R.G. Brown*; 1984–1995 W. Burger; 1984, 1989–1995 R.D. Dales (City Council representative); 1984–1995 D. de la Hey; 1984–1995 W.C. Dreboldt; 1984–1985 Mrs D. Fowle; 1984–1988 S.L. Gawith; 1984–1989 N.J. Hancock; 1984–1995 A.G. Hardingham; 1984–1988 P.C. Harwood (City Council representative); 1984–1986 T.G. Henderson*; 1984–1986 L.P. Henderson; 1984–1997 A.R. Hesp*; 1984–1986 T.E.B. Hill; 1984–1994, 1997 K.R. Howes*; 1984 D.D. Kemp; 1984–1995 D.P. Kimber; 1984–1987, 1992–1993 R.B. Lobban*; 1984 H. Lundie (City Council representative); 1984–1995 Mrs B. Merrick; 1984–1995 E.R. Merrick; 1984–1986 E.F. Mitchell-Innes; 1984–1988 E.W. Norton*; 1984–1995 M.N. Oldfield*; 1984–1988 N.R. Pinnell*; 1984–1995 T.A. Polkinghorne; 1984–1998 G.W. Poole*; 1984–1987 J.B. Poole; 1984–1988 A.P. Smith; 1984–1994 F.D. Snalam; 1984–1993 W.S. Stiller; 1984–1993 G.W. Tedder; 1984–1995 H.W. Tully; 1984–1988 Mrs M. Williams*; 1984–2013 D.A. Wing*; 1984–1988 D.R. Worrall; 1985–1986 W.J.A. Gilson (City Council representative); 1985–1994 R.F. Haswell (City Council representative); 1985–1987 Mrs F. Moyle; 1985–1988 W. Smith; 1986–1998 J.F. Plummer*; 1987–1995 G.D. de Beer (City Council representative); 1987–1995 Mrs D. Fitzsimons; 1987–1995 J.A. Richardson; 1987–1998 P.L. Smith; 1987–1995 H.D. Spencer*; 1988–1995 R.N. Barnes; 1988–1993 C.L. Greene; 1988 Mrs D.R. Roberts; 1988–1994 Dr M.J.O. Taylor*; 1989–1995 A.P. Austen Smith; 1989–1997 Mrs T.J. Hancock; 1989–1990 C.D. Harris; 1989–1995 Dr F.A.S. Hathorn; 1989–1995 B.E. Hundley; 1989–1995 Mrs M. Hurt; 1989–1990 Miss P.A. Reid; 1989–1995 Miss D. Robinson*; 1990–1994 Dr W.G.M. Galliers; 1990–1995 Dr I. Stewart; 1991–1995 W.N. MacGillivray; 1991–1992 D.W. Schofield; 1992–1995 F.A. Krause*; 1992–1995 Cmdt/Lt Col. H.D.M. Witherspoon*; 1993–1995 G.D.J. Atkinson*; 1993–1994 N.R. Bauer; 1994–1995 P. Barry; 1994–1995 I.G. Dixon; 1994–1995 Dr R.C. Nixon; 1994–1995 C. Tweedale; 1994–1995 W.D. Winship

K. Executive Committee Members

This committee included Presidents, Vice-Presidents and Honorary Life Presidents as listed above. The **elected members** were (* denotes member of the management committee. Those listed with no terminal date were still serving on the executive committee in 2021)

1996–1999, 2001–2008 G.D.J. Atkinson*; 1996–1999 J.S. Dube (City Council representative); 1996, 2001–2002 F.A. Krause; 1996–1999 W.F. Lambert (City Council representative); 1996–1997 J. Perumal (City Council representative); 1996–1998 G.W. Poole*; 1996–1998 J.M. Ronald; 1996–1998 P.L. Smith; 1996–1998 H.D. Spencer*;

1996–1998 H.R. von Klemperer; 1997–1998 J.R. Smith; 1997–2003 D.A. Wing*; 1998–2000 K. Chetty (City Council representative); 1998–1999 W.C. Dreboldt; 1999 D.H.M. Baxter*; 1999–2009 G.I. Ellis*; 1999–2002 M.J. Frickel; 1999–2006 A.J. Line*; 1999–2001 Mrs L.C. Stuart; 2000–2001 P. Breytenbach; 2001–2006 D.C. Nxumalo (City Council representative); 2001–2006 S.J. Seymour (City Council representative); 2002–2008 D. Hughes*; 2003 A. Hirsch; 2003– K. Makan*; 2003–2012 Dr I. Stewart*; 2004–2014 C.A. Froneman; 2005–2018 A. Shaw*; 2007–2017 S. Colenbrander; 2007–2011 V. Baijoo (City Council representative); 2009–2016 Ms D. Fitzsimons; 2009–2012, 2015–2018 R.J. Hex*; 2010–2018 Ms M. du Preez; 2013– M. Moncur*; 2013–2014, 2019– C. Scott; 2017–2018 J. van der Vegte; 2018– C. Dunbar; 2019– R. McDonald; 2019– T. Nixon; 2019– N. Thomas; 2019– A. Williamson

L. General Managers

1974–1994	M.G. Shute*
1994–1996	W.R. Waller
1996	Mrs J.M. van Niekerk (Acting Administrative Manager and Secretary)
1996–	T.D. Strachan (designated CEO from 2010)

M. Grounds Managers

1984–1996	D.W. (Don) Byres
1997–2001	K. Singh (Grounds Superintendent)
2002–2020	A. Motilall (Grounds Superintendent)

N. Official Guests of Honour

1984	The Honourable, the Judge-President of Natal, Judge A.J. and Mrs Milne
1985	His Honour the Administrator of Natal, Mr Radclyffe and Mrs Cadman
1986	The President of the Natal Agricultural Union, Mr Boet and Mrs Fourie
1987	The Chief of the Army, Lieutenant General and Mrs A.J. Liebenberg S.S.A.S.; S.D.
1988	The Mayor and Mayoress of Pietermaritzburg, Cllr and Mrs Mark Cornell
1989	The Honourable J.H. Steyn, Chairman of the Urban Foundation
1990	The Judge-President of Natal and Mrs Allan Howard
1991	The Minister of Agriculture and Agricultural Development the Hon. Kraai and Mrs van Niekerk
1992	Mr Christopher Saunders, Chair of the Tongaat-Hulett Group Ltd and Mrs Saunders
1993	Mr John Strong, Chair of the Zimbabwe Agricultural Society and Mrs Strong
1994	Mr John Hall, Chair of the Peace Committee
1995	Mr George Bartlett, Minister of Agriculture KZN
1996	Mr Desmond Craib, Chairman of the *Natal Witness* and Mrs Craib
1997	Dr Conrad Strauss, Group Chairman of Standard Bank
1998	Dr Ben Ngubane, the Premier of KwaZulu-Natal
1999	Dr Hermann, Executive Chairman of Clover South Africa
2000	Lord Vesty, Chairman of the Royal Agricultural Society of the Commonwealth
2001	Mr Narend Singh, KwaZulu-Natal Provincial Minister of Agriculture and Environmental Affairs

2002	Mr Alec Erwin, National Minister of Trade and Industries
2003	Dr Lionel Mtshali, the Premier of KwaZulu-Natal
2004	Mrs Nombulelo Moholi, Chief Sales and Marketing Officer, Telkom
2005	Hon. Mr Justice V.E.M. Tshabalala, Judge-President of KwaZulu-Natal
2006	Mrs Ina Cronjé, MEC for Education in KwaZulu-Natal
2007–2010	No official openings of the Royal Show and no official guests of honour recorded
2011	Dr Z. Mkhize, the Premier of KwaZulu-Natal, for the 160th Anniversary of the RAS
2012–2021	No official openings of the Royal Show and no official guests of honour recorded

O. Royal Show attendance statistics

Show duration: 1984–1992 nine days each; 1993–2019 ten days each

Year	Attendance	
1984	207 341	125th Show
1985	166 509	
1986	189 643	
1987	200 225	
1988	221 974	Pietermaritzburg 150th Anniversary
1989	214 220	
1990	206 615	
1991	212 649	Natal Agricultural Union Centenary
1992	210 466	
1993	169 647	
1994	183 624	
1995	178 225	
1996	189 203	*Natal Witness* 150th Anniversary
1997	197 757	
1998	249 270	
1999	203 627	Clover South Africa Centenary
2000	206 688	
2001	228 813	RAS 150th Anniversary
2002	207 911	
2003	222 719	
2004	214 606	
2005	225 660	
2006	161 057	
2007	170 353	
2008	155 475	
2009	131 060	150th Show
2010	136 031	
2011	146 312	RAS 160th Anniversary
2012	153 451	
2013	142 194	
2014	145 256	
2015	132 643	
2016	132 163	

2017	124 664		
2018	129 050		
2019	117 037	160th Show	
2020		Show cancelled	
2021		Show cancelled; RAS 170th Anniversary	

P. Membership

	Adult	Junior	Life	Competitor members	Total
1984	3 691	780	117		4 588
1985	3 538	739	117		4 394
1986	3 407	641	117		4 165
1987	3 376	579	123	86	4 164
1988	3 556	628	126	64	4 374
1989	3 562	652	126	70	4 410
1990	3 669	667	127	78	4 541
1991	3 629	663	128	156	4 576
1992	3 636	617	132	244	4 629
1993	3 537	627	116	264	4 544
1994	3 584	547	96	285	4 512
1995	3 424	496	96	269	4 285
1996	3 209	459	80	500	4 248
1997	3 417	477	80		3 974
1998	2 965	398	80		3 443
1999	2 906	379	71		3 356
2000	2 590	396	70		3 056
2001	2 444	398	85		2 927
2002	2 240	337	85		2 662
2003	2 123	340	81		2 544
2004	2 145	361	79		2 585
2005	2 074	325	79		2 478
2006	2 048	345	81		2 474
2007	1 942	267	88		2 297
2008	1 735	252	102		2 089
2009	1 652	303	90		2 045
2010	1 313	159	84		1 556
2011	1 296	158	81		1 535
2012	1 446	189	85		1 720
2013	1 486	160	86		1 732
2014	1 494	184	83		1 761
2015	1 470	183	90		1 743
2016	1 341	182	88		1 611
2017	1 114	122	83		1 319
2018	985	95	80		1 160
2019	1 085	99	82		1 266
2020	1 085	99	82		1 266
2021	1 013	80	74		1 167

BIBLIOGRAPHY

A.   **UNPUBLISHED SOURCES**

1.   **Royal Agricultural Society (RAS) Records**
Committee Meeting Minutes and Appendices, 1984–2019
   (a)   Annual General Meetings and Special General Meetings, General Manager's/
         CEO's Notes, Observations and Reports to Management Committee,
         Miscellaneous Memoranda and Proposals
   (b)   Section Committees: Agricultural Implements, Agricultural Services,
         Apiarian, Cattle (including Beef Cattle, Cattle Executive, Commercial
         Cattle, Dairy Cattle and Sussex Sub-Committees), Combined Chairman's
         Committee, Commerce and Industries, Dairy Produce, Forestry, Fur and
         Feather, Horse (including Horse Sub-Committee Chairmen), Pig (including
         Pig Think Tank), Poultry and Bird Clubs, Section Chairmen, Sheep, Sheep
         and Goat, Sheep Shearing
   (c)   Other Committees: Access Control, Agriculture and Produce, Arena
         Directors, Catering, Cost Saving, Dual Purpose, Executive, Finance
         and Capital Projects, Garden and Leisure, Gates, General, Grounds and
         Maintenance, Ideas, Management, Marketing, President and Vice-Presidents

2.   **Typescripts**
Royal Agricultural Society Office of the CEO. 'The Royal Agricultural Society of
Natal strategic overview' (January 2016)
Shute, Mark. 'KwaZulu-Natal's Royal Show' (15 March 2001)

3.   **Personal Information**

Cornell, Bert	Interview, 18 June 2019
Dreboldt, Cynthia	Interview, 5 March 2018
Du Preez, Marlize	Interview, 5 August 2019
Easthorpe, Ken and Mark	Interview, 27 August 2019
Ellis, G.I. (Garth)	Interview, 13 September 2019
Fitzsimons, Di	Interview, 4 June 2019
Fowler, J.M. (John)	Interview, 10 June 2019
Fraser, Jenny	E-mail, 1 July 2019
Glaister, Paula	Interview, 7 August 2019
Line, A.J. (Andrew)	Interview, 29 July 2019
Makan, Kay	Interview, 2 July 2019
McDonald, Ron	Interview, 12 June 2019
Moncur, Mike	Interview, 26 June 2019
Nixon, Tim	Interview, 1 August 2019
Poole, Moira	Interview, 1 August 2019
Spencer, H.D. (Derek)	Interview, 29 May 2019
Stewart, Dr Iona	Interviews, 16 January, 12 February and 29 October 2019
Strachan, T.D. (Terry)	Interviews, 6, 8, 12 March and 17 October 2019; E-mails, 7 October 2020 and 17 December 2020
Veness, Melanie	Interview, 11 December 2020
Von Klemperer, Margaret	E-mail, 6 July 2019

Will, Janice   E-mail, 20 May 2021, attaching Graham Atkinson
                              E-mail to T.D. Strachan, 16 May 2021
Wing, Reeva   Interview, 5 July 2019

4.   Personal Papers
Easthorpe, Ken. '62 years' (unpublished typescript)
Fowler, J.M. Press Cuttings
McDonald, R. Miscellaneous Correspondence
Spencer, H.D. Miscellaneous Correspondence, Notes and Speeches
Stewart, Dr Iona. Miscellaneous Correspondence, Notes and Speeches

B.   PUBLISHED SOURCES

1.   Royal Agricultural Society Publications
Annual Year Books, 1984–1989
Annual Reports and Financial Statements, 1984–2021

2.   Books, Journal Articles and Pamphlets
Gordon, Ruth. *Natal's Royal Show* (Pietermaritzburg: Shuter & Shooter, 1984)
Guest, Bill. *A Century of Science and Service: The Natal Museum in a Changing
      South Africa 1904–2004* (Pietermaritzburg: Natal Museum, 2006)
Guest, Bill. *A Fine Band of Farmers Are We! A History of Agricultural Studies in
      Pietermaritzburg 1934–2009* (Pietermaritzburg: Natal Society Foundation, 2010)
Guest, Bill. *Trek and Transition: A History of the Msunduzi and Ncome Museums
      (incorporating the Voortrekker Museum Complex) 1912–2012* (Pietermaritzburg:
      Msunduzi and Ncome Museums, 2012)
Guest, Bill. *Stella Aurorae: The History of a South African University Volume 3 The
      University of Natal (1977–2003)* (Pietermaritzburg: Natal Society Foundation,
      2019)
Hesp, Tony. 'Hubert Ralph von Klemperer (1914–1999)' *Natalia* 29 (1999): 102–105
Laband, J. and Haswell, R. (eds). *Pietermaritzburg 1838–1988: A New Portrait
      of an African City* (Pietermaritzburg: University of Natal Press and Shuter &
      Shooter, 1988)
The Royal Agricultural Society. *Prospectus 2012–2015* (Pietermaritzburg: RAS,
      2015)
Young, Lindsay. *A History of the Royal Agricultural Society of Natal, 1851–1953*
      (Pietermaritzburg: RAS, 1953)

3.   Newspapers and Periodicals (selectively)
Farmer's Weekly
Maritzburg Fever
Natal Witness/Witness
Public Eye, March 2013
Stud Breeder, April 2011
Weekend Witness

4.   Royal Agricultural Society Press Releases and Newspaper Clippings
1851–1988, 1994, 2013–2021

5.   Websites
https://en.wikipedia.org/wiki/Rand_Show
https://www.tanya.visser.com

INDEX

Page numbers in **bold** indicate both illustrations and text

www.ingramcontent.com/pod-product-compliance
Lightning Source LLC
Chambersburg PA
CBHW080356030426
42334CB00024B/2897